EMBEDDED ENTERPRISE AND SOCIAL CAPITAL

Embedded Enterprise and Social Capital

International perspectives

Edited by

MICHAEL TAYLOR
University of Birmingham

SIMON LEONARD
University of Portsmouth

Routledge
Taylor & Francis Group

LONDON AND NEW YORK

First published 2002 by Ashgate Publishing

Reissued 2018 by Routledge
2 Park Square, Milton Park, Abingdon, Oxon OX14 4RN
711 Third Avenue, New York, NY 10017, USA

Routledge is an imprint of the Taylor & Francis Group, an informa business

Publisher's Note
The publisher has gone to great lengths to ensure the quality of this reprint but points out that some imperfections in the original copies may be apparent.

Disclaimer
The publisher has made every effort to trace copyright holders and welcomes correspondence from those they have been unable to contact.

A Library of Congress record exists under LC control number: 2002100898

ISBN 13: 978-1-138-73438-8 (hbk)
ISBN 13: 978-1-138-73437-1 (pbk)
ISBN 13: 978-1-315-18683-2 (ebk)

Contents

List of Figures

List of Tables

List of Contributors

Pierre Agnes, Lecturer, School of Marketing, The University of New South Wales, Sydney, New South Wales, Australia.

Ron Boschma, Associate Professor, Department of International Economics and Economic Geography, The University of Utrecht, Utrecht, The Netherlands.

Carol Ekinsmyth, Senior Lecturer in Geography, Department of Geography, University of Portsmouth, Portsmouth, UK.

Ayda Eraydin, Professor, Department of Urban and Regional Planning, The Middle East Technical University, Ankara, Turkey.

Irene Hardill, Professor of Economic Geography, School of International Studies, Nottingham Trent University, Nottingham, UK.

David Hayward, Senior Lecturer, Department of Geography, The University of Auckland, Auckland, New Zealand.

Jan Lambooy, Professor of Regional Economics, Department of International Economics and Economic Geography, The University of Utrecht, Utrecht, The Netherlands.

Richard Le Heron, Professor of Geography, Department of Geography, The University of Auckland, Auckland, New Zealand.

Simon Leonard, Senior Lecturer in Geography, Department of Geography, University of Portsmouth, Portsmouth, UK.

Giles Mohan, Senior Lecturer in Development Studies, The Open University, Milton Keynes, UK.

Sara Openshaw, Best Value Review Manager, Dorset Police, Dorchester, Dorset, UK.

Montserrat Pallares-Barbera, Asssociate Professor, Department of Geography, Universitat Autònoma de Barcelona, Barcelona, Spain.

Parvati Raghuram, Lecturer, Department of International Studies, Nottingham Trent University, Nottingham, UK.

Izhak Schnell, Associate Professor, Department of Geography, Tel Aviv University, Tel Aviv, Israel.

Veronique A.J.M. Schutjens, Lecturer and Researcher, The Department of International Economics and Economic Geography, University of Utrecht, Utrecht, The Netherlands.

Paul Search, Risk Consultant, Marsh Ltd, London, UK.

Michael Sofer,   Head of Geography at the Levinski Teachers' College and Lecturer in the Departments of Geography at Tel-Aviv and Bar-Ilan Universities, Israel.

Adam Strange, Administrator, The Open University, Milton Keynes, UK.

Christina Stringer, Senior Tutor, Department of Geography, The University of Auckland, Auckland, New Zealand.

Michael Taylor, Professor of Human Geography, School of Geography and Environmental Sciences, The University of Birmingham, Birmingham, UK.

Kjersti Wølneberg, Higher Executive Officer, Norwegian Ministry of Education and Research, Oslo, Norway.

Chapter 1

Approaching 'Embeddedness'

Michael Taylor and Simon Leonard

Introduction

In the past decade, a powerful model of local economic growth has developed that derives in the first instance from Granovetter's (1985) concept of 'embeddedness'. The model draws on a range of complementary literatures on 'new industrial spaces', 'clusters', 'learning regions' and 'innovative milieu' and 'regional innovation sytems' (Braczyk *et al.* 1998, Maskell *et al.* 1998, Oinas 1997, Porter 1998, Simmie 1997, Storper 1997, Taylor and Conti 1997, Bergman *et al.* 2001). The model emphasises the social construction of inter-firm relationships, collaborative supplier-buyer interaction, the creation of 'social capital' and the local significance of 'institutional thickness'. It has obvious policy relevance and appeal, and 'social capital' is now increasingly being used as a catch-all phrase in political economy to cover the non-economic and non-political relations which underpin successful economic development and forge sustainable democracies (Maskell 1999, Putnam 1993). Clustering concepts, in particular, are being used as the foundation of local economic development policies in a large number of OECD countries (Department of Trade and Industry 2001, Bergman *et al.* 2001).

In the model, successful local economies are recognised as islands of economic activity and sustained local accumulation. Built on superior local productivity (Porter 1998), and integrated into a global mosaic of production. Success and growth in these local economies is argued to be dependent upon complex processes of 'embedding' that involve trust, reciprocity, loyalty, collaboration and cooperation rooted in a place. These are processes creating social capital. They generate information, ideas and innovation and are exchanged through mechanisms of quasi-integration involving collaboration and sharing rather than appropriation (Leborgne and Lipietz 1992). The whole is a process of collectivisation that creates 'institutional thickness' which

1

further bolsters this local economic process (Powell and Di Maggio 1991, Amin and Thrift 1994a, 1995, 1997). The global mosaic is, in turn, orchestrated by TNCs, global capital and global political institutions (e.g. IMF, WTO etc.). Enmeshed in webs of global coordination and value transfer, local economies are interpreted in this model as nodes of un-traded dependencies that are tapped into by internationalised and globalised corporations. The corporations act as global information arbitrageurs (*The Economist* 1995) in an era of networked 'soft capitalism' (Thrift 1998).

Embeddedness is recognised as having four basic forms – cognitive, cultural, political, and structural (Zukin and Di Maggio 1990, Grabher 1993a). *Cognitive embeddedness* identifies the bounded rationality of economic actors and place-based knowledge. *Cultural embeddedness* recognises the importance of shared collective understandings in decision-making and goal formulation amongst firms doing business in a place. *Political embeddedness* recognises the place-based impact on firms' business decisions of struggles with non-market institutions that might just as easily foster them (Heidenreich and Krauss 1998) or constrain them to the point of failure (Glasmeier 1991, Grabher 1993b).

At the heart of the embeddeness thesis, however, is *structural embeddedness* which identifies the manner in which business enterprises are incorporated into local, place-based networks that facilitate and promote information exchange and learning (Maskell *et al.* 1998, Asheim 1996). It is argued here that there are significant aspects of structural embeddedness that have been either omitted or underplayed in recent analyses of 'clustering' and new industrial spaces. In particular, the role and impact of power asymmetries between business enterprises remains largely unexplored in this literature despite the central role it has been identified as playing in network relationships (see Grabher 1993a, Taylor 2000).

Structural embeddedness has been recognised as having four essential characteristics; *reciprocity, interdependence, loose couplings* and *asymmetric power relations* (Grabher 1993a, pp. 8-12). *Reciprocity* refers to recurrent transactions between networked firms that are more than simply repetitive and involve relationships that do not have immediate equivalence in each transaction but achieve some approximate balance over the life of an exchange relationship (Polanyi's (1957) 'gift economy'). *Interdependence* reflects the elements of trust and mutual orientation in long-term exchange relationships that enable firms to exchange resources and information that are crucial for high performance but are difficult to value and transfer via market ties (Uzzi 1996, p. 678). It is central to network learning and local innovation capacities. *Loose couplings*, or integrated separateness (Lundvall 1993), recognises the ability of firms networked in a place individually to shift their partners while maintaining an essentially stable district framework of interaction. *Asymmetric power relations* are a counterweight to the cosiness of network collaboration, with collaboration and cooperation within networks being undermined by practices of dominance and exploitation between unequal exchange partners (Dicken and Thrift 1992, Grabher 1993b, Taylor 2000).

Much of the most recent literature on industrial districts, learning regions, regional innovation systems and innovative milieu has all but ignored issues of asymmetric power relations and their impact on network relationships and the nature of exchange and transaction structures in local economies (see for example the separate contributions in Braczyk *et al.* 1998; Conti *et al.* 1995, Bergman *et al.* 2001, Taylor and Conti 1997). Certainly, cooperation and collaboration in local economies have been recognised as involving important dynamic tensions, but these tensions have been seen arising essentially from competition and not from the more brutal exercise of power between firms (Brusco 1996, Porter 1998).

Cooperation, it is argued, must be tempered with competition. Some inter-firm transactions in local economies might best be undertaken collaboratively and cooperatively while others might be best left to price determined competition (Cooke 1998, Enright 1995). A balance is needed, therefore, to avoid firm lock-ins, to avoid paternalism in the labour market; to stimulate the knowledge economy; and to prevent 'institutional overload' obscuring economic imperatives (Uzzi 1996, 1997).

However, the concept of 'embeddedness' is imprecise and ill-defined. In the model of local growth it has spawned in geography, sociology and some areas of management science, the concept is undeniably vague, but it has directed attention to the nature of relationships between firms and their socio-spatial environments that are neither well understood nor particularly well conceptualised (Oinas 1997). The assumption implicit in the model is, however, that 'embedded' equals 'local'. Oinas (1997, p. 29) has argued that there is no reason for this to be so and that entrepreneurs and business people can be embedded in social relations at different spatial scales (see also Taylor *et al.* 1997 on 'conflicting spatialities'). Nevertheless, there is no clear understanding of what aspects of social relations lead to the cultural and political embeddedness of firms. On the vagueness of the concept, Oinas (1997) has remarked on the need to understand:

(1) the various ways in which firms ... are embedded, and the ways in which these different embeddednesses are related to each other and to economic outcomes ... (2) the degree to which firms' embeddedness in local social relations enhances or hinders processes of change in both firms and their local

environments ...(3) the degree to which extra-local embeddedness of firms encourages economic development in some cases, and inhibits it in others. (p. 30)

Indeed, the expanding literature on commodity chains suggests that 'proximity' and 'locational integration', that are fundamental to the local embeddedness model, may not be the only mechanisms that can promote enterprise 'learning' and the creation of local social capital. A global commodity chain is, 'a set of transnational organisational linkages that constitute the production, distribution and consumption of a commodity' (Korzeniewicz 1992, p. 314). They are an expression of the managerial coordination that time space compression affords (Gereffi 1994, 1999). Notwithstanding the power asymmetries in these commodity chains, it has been argued that in buyer-driven chains there is the potential for knowledge transfer and 'learning' (not dissimilar to processes identified in clusters) that can foster 'industrial upgrading' in the lower levels of the commodity chain (Gereffi 1999, Hsing 1999). This upgrading can involve firms engaging in product elaboration and a shift to complex, expensive, large volume, high end products; shifting to flexible production; beginning original equipment manufacturing (OEM) and original brand manufacturing (OBM); and, at a regional scale, developing a locally integrated production system (Gereffi 1999, pp. 51-52). Whether the world is always so simple is open to debate, but the suggestion is quite clear that 'learning' in economic systems and the creation of social capital need not always involve proximity.

Furthermore, the model, built on the concept of embeddedness in geography and development studies, is now increasingly being criticised. That criticism is challenging the nature and extent of local embeddedness as the foundation of local economic growth. As part of this critique, the symmetrical properties of trust, reciprocity and loyalty in buyer-supplier relationships are argued as being either temporary or even illusory, and to be fundamentally at odds with the existence and impact of power asymmetries within and between firms (Bresnen 1996, Pratt 1997, Baker 1996, Taylor 2000). The model is seen by some as an a-historic, idealised and romanticised view of inter-firm relationships that inappropriately extends notions of flexibility, is policy driven, functionalist and is perpetuated by selecting case studies 'on the independent variable' (Bianchi 1998, Hudson 1999, Staber 1996, Lovering 1999).

Against the background of this emerging critique, what is needed is a fuller, deeper and nuanced understanding of: (1) the nature of inter-firm and enterprise/institution relationships; (2) the relationship between 'structural', 'cognitive', 'cultural' and 'political' embeddedness; (3) the inclusionary and exclusionary tendencies of embeddedness; (4) the relationship between firm embeddedness and labour market conditions; (5) the role of TNCs in linking global and local networks; (6) the interplay of power and reciprocity in complex economic systems, and the processes of institutionalisation as they shape growth; and, (7) the spatialities of embeddedness.

Embeddedness in Context

In the extensive literature on small enterprise spatial systems (new industrial spaces, learning regions, economic milieu, regional innovation systems and clustering), three quite distinct 'contexts' of embeddedness can be recognised which are essentially variants of the embeddedness model of local economic growth. First, there is the developed country context , within which 'inclusive' embeddedness is seen as a mechanism for (re)generating international competitiveness in mature industrial economies. Second, there is the developing country context in which 'inclusive' embeddedness is seen as a mechanism and strategy for coping locally with the pressures and problems of economic globalisation. Third, there is the TNC context in which 'inclusive' embeddedness is a globalisation strategy for tapping into new economic spaces.

In each of these contexts the emphasis in current theoretical literature is on the incorporation of individual firms into networks of reciprocal exchange, in local production chains, as a prerequisite for incorporation into the global economy. The route to international competitiveness is through the dynamic inclusiveness of weak ties. What have been neglected, if not ignored, however, have been the countervailing processes of exclusion created not just by the weakness of strong ties but also by the strength of strong ties.

In the context of local growth in *mature industrial economies* there is a massive literature on clustering, learning, milieu and un-traded dependencies that highlights Schumpeterian innovation, knowledge transfer and the creation of 'atmosphere' as the positive consequences of structural embeddedness (Asheim 1996, 2000). And, when these supply chain mechanisms are coupled with enhanced productivity they supplement or maintain the global competitiveness of a place. It is widely contended in this literature that in a new era of flexible specialisation and flexible accumulation (Piore and Sabel 1984, Sabel 1989), economic growth and the survival of business enterprises at the local scale is dependent upon the incorporation of firms into socially embedded networks of collaborative production (Cooke, 1998), '... buttressed by a supportive tissue of local institutions' (Powell and Smith-Doerr 1994, p. 370). The benefits of these collaborative and cooperative relationships, that are as much social as economic, are seen in terms of heightened place-based capacities for learning, information and knowledge exchange, technological change and innovation amongst network members (Maskell and Malmberg 1999, Bergman *et al.* 2001). Industrial districts are, by this interpretation, regional innovation systems and engines of local economic growth and competitive strength fuelled by the intensified and localised processes of Schumpeterian creative destruction. In this way, local economies remain nimble in the face of processes of 'ubiquification' which leave firms constantly faced by the need to cope with shifting factor mixes (Maskell *et al.* 1998). Breathing Marshallian 'industrial atmosphere', these firms are incorporated as dynamic nodes of innovation and learning into an emerging global mosaic of regions (Scott and Storper 1992, Storper 1997, Asheim 2000).

But, what are ignored in these essentially functionalist interpretations of dynamic localities in mature industrial economies are the countervailing processes of exclusion. Already it has been recognised that embeddedness cannot replace market mechanisms entirely (see for example Uzzi 1996, 1997). Cooperation, as stated earlier, must be tempered by competition (Cooke 1998, Enright 1995). The case can be argued, however, that economic relationships involve more than a dynamic tension between collaboration and competition, they also involve the more brutal exercise of power, through the control of resources, the manipulation of relationships or the exercise of discipline (Taylor 1995, Allen 1997). Power inequalities can be seen leading to exclusion in at least four ways. First, they restrict firms' freedom of action. Power inequalities limit the forms of transaction and buyer-supplier relationships open to firms according to their positions in inter-firm networks (Taylor 1995, 2000, Hallsworth and Taylor 1996). This restricts the ways they are able to do business and, thus, their possibilities and potential to accumulate capital. Second, power inequalities can create lock-ins and the ossification of transaction relationships – a process that has been well detailed by Amin and Robins (1990), Amin (1993), Grabher (1993b) and Glasmeier (1991). Indeed, the creation of institutions in a place can be seen just as much as a way of protecting the status quo of doing business in a place (and so promoting lock-in) as it is a mechanism for the generation of dynamic development. Third, power inequalities lead to uneven spatial development – a well recognised inherent characteristic of capitalist accumulation (Hudson 1999, Massey 1984). Just as some local economic systems are 'winners', others are 'losers'. Finally, the power inequalities of class lead to exclusion in business communities through its shaping of boards of directors and the strategic decision-making role they have. (Hambrick and Mason 1984, McNulty and Pettigrew 1999). However, the arguments that counter the embedded local growth model are as yet only weakly developed in the developed country context.

While the local embeddedness model has been accused of being policy-driven theory in the context of the industrial west (Lovering 1999), it is very clearly seen as a way of promoting socially and culturally sensitive economic growth in the *developing country context*. A range of analysts have suggested that the embeddedness processes at the heart of the model can be translated into policies whereby regions and localities attempt to accrete social capital and build institutional thickness (Cooke 1998, Amin and Thrift 1997, Putnam 1993). Here though, there is a major implicit shift in the focus of the embeddedness model. Though first recognised as a competitive strategy for the incorporation of places in a global economy, here it is subtlely translated into a coping mechanism to counter the underdevelopment of places by global economic pressures. However, while analysts generally report examples where institutions have been part of economic success, they are less certain how to build these social relations from scratch. The problem is that these partial analyses reify certain institutional ensembles, elevating them to the status of universal mechanisms, while at the same time cementing key stories in the intellectual-political consciousness (Staber 1996, Bianchi 1998). There is now the great danger that the mere existence of certain institutions is sufficient proof of their economic effectiveness without actually demonstrating the causal linkage between their actions and economic outcomes.

Now in the developing country context the focus is on the creation of 'social capital' and the creation of seemingly appropriate non-economic and non-political institutions as the underpinnings of successful economic growth (Putnam 1993). Such institutionally driven research is relatively new (Schmitz and Musyck 1993, Honig 1998). The major lending institutions, however, have recently been pushing for 'good governance' to create an 'enabling environment' for economic growth in the developing world, so it can be assumed that such studies will be of growing importance as these policy initiatives mature (Mohan 1996).

Tentative conclusions from empirical research show, however, that in developing countries where scale economies are lacking, there is a greater need for inter-firm collaboration to foster local economic growth (Rasmussen *et al.* 1992). However, a study by Schmitz (1993) on Brazil showed that collaboration and reciprocity based on a shared cultural heritage of migrants broke down when competition from other industrial districts – in this case Korean manufacturers – forced firms to interact largely on a cost/price basis. It also showed that social ties between manufacturers actually permitted collusion and the concentration of power within a small number of large firms. The results for their subcontractors were far less autonomy and increased hardship for their workers. Hence, the success of this Brazilian industrial district was predicated only weakly upon processes of social and institutional cooperation that took them far from the 'high road' of economic development envisioned by the regional boosters (Ohmae 1995, Cooke 1998). What we do not know is how extensive such *exclusionary* processes are in the global periphery, and we certainly have no strong evidence to assume they do not exist in the developed country context as well. Indeed, there is an urgent need to unlock the complex and specific local relations between individuals, firms and institutions under different conditions of economic development.

In the context of TNCs moving into new markets and new economic environments, embeddedness is again a strategy to enhance international competitiveness built on inclusion and incorporation into specific local contexts. In the process of becoming transnational, TNCs grow out of particular local contexts and acquire characteristics of those places through the complex processes of networking and embedding that create them (Dicken and Thrift 1992, Dicken 1994, Dicken *et al.* 1994). They carry that local flavour with them as they move into the international arena (Yeung 1998a, p. 116). However, what might be successful in their home place might not be successful in another where there are significant place-specific differences. Those differences might include, for example: 'Gentlemen's' agreements on the way business is done; the manner and mechanisms of negotiating with other businesses, governments and political elites; the understandings developed through personal contacts; the accumulation of intelligence, information and knowledge; processes of informal recommendation and the building of reputations; and the nature of local buyer-supplier relationships and the foibles of local markets (see Crew 1996, Yeung 1998b).

Such place-specific knowledge, it is argued, is only gained from local experience; so foreign firms have to make themselves less 'foreign'. They have to be embedded in the networks of the host countries they operate in so that simultaneously they can be globally integrated and yet locally responsive. Indeed, Yeung (1998a) draws on the work of, amongst others, Pryke and Lee (1995), Clark (1997) and Amin and Graham (1997) to argue that the external and territorial economies of financial production in centres such as the City of London, New York and Tokyo that derive from complex networks of social and personal relationships, can only be exploited by transnational financial institutions when they are locally embedded in those places (p. 301). Jonas (1996) has suggested the same local embedding of US TNCs in Mexico's northern border towns, and Fujita and Hill (1995) have outlined Japanese TNCs' local embedded relationships with firms and institutions in the USA. In a very different context, Yeung (1997a, 1997b) has considered the success of Chinese business organisations to derive from their embeddedness in distinctively local business networks involving a, 'perpetual tendency to cultivate complex networks of personal and business relationships that are rooted locally among Overseas Chinese Diaspora' (Yeung, 1998b, p. 119). Furthermore, it is suggested that these essentially spatial networks offer participants formidable first-mover competitive advantages.

The message is clear, the inclusiveness of the embeddedness model embraces large corporations as much as it does SMEs, and by being reciprocating network members TNCs can prosper locally and be internationally competitive as part of a corporate globalisation strategy. This is a very different story to the long-standing discourse that highlights the exploitative, exclusionary view of TNCs, as expressed in Dunning's (1979) OLI model, the New International Division of Labour thesis (Fröbel *et al.* 1980), radical perspectives on the geographical transfer of value (Forbes and Rimmer 1984), or the extensive critique of US transnational corporations developed by, amongst others, Barnett and Müller (1978) and Bergsten *et al.* (1978). It would seem important, therefore, to begin to counterbalance the inclusionary tendencies inherent in the local embeddedness of TNCs with the exclusionary tendencies that stem from their exercise of power in those same places (see Allen and Pryke 1994).

These three contexts of embeddedness, and their different tendencies towards the local inclusion and exclusion of business enterprises, reinforce the relevance of Yeung's (1998b) proposed agenda for research on enterprise embeddedness. At the same time, they extend it. Yeung's first agenda item is, 'to untangle the complex ways in which business organizations are locally embedded' (p. 120). But, to examine the social and institutional contexts of embeddedness and the ways, '... business organisations are constituted *in situ*' (p. 121) is to go only part way. Research also needs to address not just the mechanisms, processes and circumstances of *inclusion* in local social networks, but also the other side of the duality, the mechanisms, processes and circumstances of *exclusion*. It is just as important to know why firms and enterprises are not embedded in local systems of reciprocal relations and yet continue to prosper. For the second agenda item, exploring the '... elements that contribute to the "institutional thickness" arising from local agglomerations' (p. 121), the same point holds true. 'Institutional thickness' is by itself a profoundly unhelpful term, but it is just as important to know how it can operate to *exclude* firms as it is to know how it can *include* them and enhance their growth. Certainly the rules of meaning and membership institutionalised in places (Callon 1986) are just as exclusionary as they are facilitative (Taylor 1995). Yeung's final agenda item is to develop an understanding of the impact of enterprise power on local embeddedness. This, it can be argued is the single most important task facing embeddedness research (Taylor 1995, 2000). As the control of resources, the manipulation of relationships or the exercise of power as discipline (Clegg 1989, 1990), power within and between enterprises and institutions again has the ability to include or exclude individual actors (Pfeffer 1981, Wrong, 1995).

To fully understand the significance of local processes of embedding requires empirical assessments of the nature of the embeddedness of firms and enterprises and the associated tissue of institutions (supportive or otherwise) in a range of national and cultural settings. It is proposed here that drawing together a wide, international range of studies on different aspects of the complex processes of embedding is a first step in developing a fuller understanding of the role of networked social relationships in shaping local economic growth in developed and developing countries. By examining processes of embedding in different national and cultural settings it will be possible to begin to assess whether the model as a whole and the policy initiatives it has spawned can properly be universalised.

The Structure of the Volume

Against the background of the 'embeddedness' model of local economic development and the critique that is emerging, this volume brings together researchers who explore a wide range of issues related to enterprise embeddedness. This introductory chapter has reviewed the broad base of views on embeddedness and local economic growth that has developed in the past decade. Chapters 2 and 3 extend this discussion in very specific ways. In Chapter 2, (*'Embeddedness and Innovation'*) Boschma, Lambooy and Schutjens further elaborate the contradiction and lack of clarity that runs through the embeddedness literature, especially the apparent contradiction between the processes of learning and lock-in that can co-exist in local clusters. In Chapter 3 (*'Rethinking Institutions and Embeddedness in a Third World Context'*), Mohan highlights the inappropriateness of the model for understanding and interpreting local economic growth in Third World countries. He complains that it undervalues national policies, mis-specifies the roles of institutions, fosters anti-economism and is blind to grinding poverty.

Chapters 4 and 5 extend the critique of the embeddedness model using case study material from Argentina and New Zealand to demonstrate the impact of the unequal power relationships between firms and institutions, in buyer-driven commodity chains, on local growth capacities. In Chapter 4 (*'Supply Chains, Embeddedness and the Restructuring of Argentina's Tanning Industry'*), Wølneberg demonstrates these processes in Argentina, highlighting the importance of national and international governance structures in maintaining inequality and fostering the importance of corporate 'bigness'. Power inequalities are shown to massively outweigh collaboration as drivers of economic growth. Hayward, Stringer and Le Heron continue this theme in Chapter 5 (*'Going Places? Reflections on Embedding and Disembedding in Agriculture and Horticulture Under Neo-Liberalism: The Example of Hawke's Bay, New Zealand'*), seeing local businesses 'thickly constrained' by wider corporate and institutional networks. Constraint goes hand-in-hand with opportunity, but this New Zealand example shows how the power of just a few 'gatekeepers' can shape that constraint, guided by an organisational proximity that may have little to do with physical proximity and the interests of a locality.

Chapters 6 and 7 develop the critique of the embeddeness model from a different perspective – the time-specific processes of new firm formation and local innovation. In Chapter 6 (*'Weakening Ties: The Embeddedness of Small UK Electronics Firms'*), Openshaw and Taylor interpret new firm formation in the electronics cluster on the South Coast of the UK as time-specific, time-dependent and driven by now-defunct policy initiatives. As defence procurement policies have changed in the UK, embedded relationships in the electronics industry have withered as firms have become tied into supply chains driven principally by price and international competition. The marketisation of inter-firm relationships, coupled with strong managerialism, has emphemeralised and marginalised not only small subcontracting firms but also labour. In Chapter 7, (*'Enterprise Embeddedness and Industrial Innovation in Spain: An Overview'*), Pallares-Barbera reviews the apparent processes involved in non-metropolitan industrial growth in Spain against the postulated processes of the embeddedness model. Again, the discussion focuses on the creation and maintenance of growth among firms in local economies and the limitations of the fundamentally a-temporal theoretical model. A series of questions on the mechanisms postulated in theory arise from the analysis including: how are embedded inter-firm relationships initiated in a local cluster; how is innovation created and spread; and how does government policy fit with these processes?

Chapters 8 and 9 extend the analysis of embeddedness to cover firms in service industries – financial services in Sydney, Australia, and business services in parts of the UK. In Chapter 8 (*'Local Embeddedness in Global Financial Services: Australian Evidence on "The End of Geography"'*), Agnes maintains that it is not necessarily the case that financial services are increasingly being tied into a seamless global network. Instead, he argues from an analysis of swaps dealing, futures trading and master custody that local embeddedness is still important, but assumes different forms depending on the specifics of providing a particular service. In Chapter 9 (*'Local Embeddedness and Service Firms: Evidence from Southern England'*), Search and Taylor reinforce this argument by shifting the analysis to cover small business service firms, particularly accountants, surveyors and solicitors. Small firms providing these services are found to have strong personal ties with clients and suppliers, involving trust, reciprocity and loyalty. These embedded ties, in turn, play a key role in new firm formation through the advice and active participation of these business service providers as investors and business angels. Simultaneously, then, these service firms are the glue and the drivers of local business communities. This situation contrasts with the disembedded nature of corporate business services driven by market mechanisms and managerialism, and holding the potential to erode local processes of new firm formation.

From a different perspective again, Chapters 10 and 11 explore labour market dimensions of local enterprise embeddedness adding a new and very different dimension to the critique of the model. In Chapter 10 (*'Embedded Project-Production in Magazine Publishing: A Case of Self-Exploitation?'*), Ekinsmyth details the less benevolent aspects of embedded place-based relationships from the perspective of labour. Using evidence from magazine publishing, which is at the fore in the new 'cultural' industries, she shows how the production of magazines as 'projects' exploits 'friendship' and embedded personal relationships among freelances to create corporate profits. Of necessity, freelances must be part of local networks of reciprocity to gain and maintain work. Proximity and personal ties maintains their powerless and precarious foothold in this world of work in which they take self-exploitation back into their families and the places they live. In Chapter 11 (*'Local Embeddedness, Institutional Thickness and the State Regulation of Local Labour Markets'*), Leonard uses London to examine the impact of labour market deregulation on processes of embeddedness. The analysis of this chapter addresses the changing demands made of labour market programmes – in this instance skills training programmes – to demonstrate the interplay between government policies and the institutions they create. The chapter demonstrates the shifting and sometimes contradictory roles these institutions are mandated to perform. Through its strong focus on time and change, the chapter adds a richness and realism to the arguably naïve views on 'institutional thickness' and the role of a supportive tissue of institutions that plays a central role within the concept of embedded local growth.

Through analyses of the mechanisms shaping and limiting the success of ethnic minority enterprise in the UK and Israel, Chapters 12 and 13 shed additional light on the link between embeddedness and local growth, by highlighting the constraints imposed by societies and politics. In Chapter 12 (*'Diasporic Embeddedness and Asian Women Entrepreneurs in the UK'*), Hardill, Raghuram and Strange detail the nature of embeddedness and kinship that underpins the ventures of Asian businesswomen in Britain, highlighting its distinctive 'glocal' characteristics. Embedded relationships are found in networks that do not necessarily involve proximity, and involve imagined relationships with a more distant 'home'. These links, together with the size and well-being of the local ethnic population are shown to be the vital drivers of production and consumption practices in such communities. In Chapter 13 (*'Over- and Under-Embeddedness: Failures in Developing Mixed Embeddedness Among Israeli Arab Entrepreneurs'*), Sofer and Schnell demonstrate the compounding problems of exclusion that inhibit Israeli Arab business. Here, as elsewhere, businesses develop from local support networks that are strongly community and family based. But, their lack of access to the national and political Jewish elite, their powerlessness in business transactions outside their own community, and their reluctance to use institutional support mechanisms, leaves them 'over-embedded' in the intra-ethnic Arab milieu. Here then is embeddedness as a mechanism for coping with exclusion, not as a generator of local social capital and growth. Here it is an indicator of a vicious circle of exclusion.

These themes of exclusion, institutions and powerlessness are further elaborated in Chapters 14 and 15 that explore embeddedness in two developing country contexts – Fiji in the Pacific and Turkey in Europe. In Chapter 14 (*'Enterprise Embeddedness and Exclusion: Business Relationships in a Small Island Developing Economy'*), Taylor identifies the 'selective embeddedness' of businesses in racially divided Fiji that is simultaneously inclusionary and exclusionary. TNCs display embedded relationships, but only with the largest of local businesses in an effort to minimise commercial risk, and in so doing marginalise and exclude those small local firms where theory would suggest local social capital might be generated. Large Indo-Fijian businesses, in contrast, are strongly politically embedded to gain access to politically guided commercial opportunities bred by political cronyism. The embeddedness of the remaining small livelihood and micro firms is no more than a coping mechanism for survival in a business environment of compromised quality and graft, driven solely by price. In Chapter 15, (*'The Local Embeddedness of Firms in Turkish Industrial Districts: The Changing Roles of Networks in Local Development'*), Eraydin suggests, based on Turkish evidence, that locally embedded relationships and a supportive social milieu may be important in initiating and sustaining local economic growth but they do not guarantee international competitiveness even when they involve information exchange and 'learning'. Government policies on business clustering may create apparently embedded business relationships, and there may even be extensive institutional support. However, in the face of the economic crisis that has confronted Turkish businesses, these mechanisms have offered little or no resilience, and businesses have reverted to cost minimisation strategies to survive. Here again, when the realities of time are added to the mechanisms of the a-temporal embeddedness model, this influential model is found wanting.

The separate contributions of this volume significantly advance the emerging critique of the embeddedness model of local growth. However, there is a synergy in these contributions that Chapter 16 (*'Understanding Embeddedness'*) begins to elaborate. This elaboration includes: power, inequalities, exploitation, exclusion and coping; time, change and the dynamics of places and communities; labour processes and self-exploitation; and government, governance and control. It leads to a clarification and reiteration of Amin and Thrift's (1994b) conclusion that in trying to understand growth and change the social and the economic are intertwined processes that are impossible to separate.

References

Allen, J. (1997), 'Economies of Power and Space', in R. Lee and J. Wills (eds), *Geographies of Economies*, Arnold, London, pp. 59-70.

Allen, J. and Pryke, M. (1994), 'The Production of Service Space', *Environment and Planning D: Society and Space*, vol. 12, pp. 453-475.

Amin, A. (1993), 'The globalization of the economy: an erosion of regional networks', in G. Grabher (ed), *The Embedded Firm: On the Socioeconomics of Industrial Networks*, Routledge, London, pp. 278-295.

Amin, A. and Graham, S. (1997), 'The ordinary city', *Transactions of the Institute of British Geographers, New Series*, vol. 22, pp. 411-429.

Amin, A. and Robins, K. (1990), 'Industrial districts and regional development: limits and possibilities', in F. Pyke, W. Sengenberger and G. Becattini (eds), *Industrial Districts and Inter-Firm Cooperation in Italy*, International Institute of Labour Studies, Geneva, pp. 185-220.

Amin, A. and Thrift, N.J. (1994a), 'Living in the global', in A. Amin and N. Thrift (eds), *Globalization, Institutions and Regional Development in Europe*, Oxford University Press, Oxford, pp. 1-22.

Amin, A. and Thrift, N.J. (1994b), 'Holding down the global', in A. Amin and N. Thrift (eds), *Globalization, Institutions and Regional Development in Europe*, Oxford University Press, Oxford, pp. 257-260.

Amin, A. and Thrift, N.J. (1995), 'Globalisation, institutional thickness and the local economy', in P. Healey, S. Cameron, S. Davoudi, S. Graham and A. Madani-Pour (eds), *Managing Cities: The New Urban Context*, John Wiley, London, pp. 91-108.

Amin, A. and Thrift, N.J. (1997), 'Globalization, socio-economics, territoriality', in R. Lee and J. Wills (eds), *Geographies of Economies*, Arnold, London, pp. 147-157.

Asheim, B.T. (1996), 'Industrial districts as "learning regions": A condition for prosperity', *European Planning Studies*, vol. 4, no. 4, pp. 379-397.

Asheim, B.T. (2000), 'Industrial districts: The contributions of Marshall and beyond', in G.L. Clark, M.P. Feldman and M.S. Gertler (eds), *The Oxford Handbook of Economic Geography*, Oxford University Press, Oxford, pp. 413-431.

Baker, P. (1996), 'Spatial outcomes of capital restructuring: "new industrial spaces" as a symptom of crisis, not solution', *Review of Political Economy*, vol. 8, no. 3, pp. 263-278.

Barnett, R. and Müller, R. (1978), *Global Reach: The Power of the Multinational*, Jonathan Cape, London.

Bergman, E.M., Charles, D. and den Hertog, P. (eds) (2001), *Innovative Clusters: Drivers of National Innovation Systems*, OECD, Paris.

Bergsten, C., Horst, T. and Moran, T. (1978), *American Multinationals and American Interests*, The Brookings Institute, Washington DC.

Bianchi, G. (1998), 'Requiem for the third Italy? Rise and fall of a too successful concept', *Entrepreneurship and Regional Development*, vol. 10, pp. 93-116.

Braczyk, H.J., Cooke, P. and Heidenreich, M. (eds) (1998), *Regional Innovation Systems. The Role of Governances in a Globalized World*, UCL Press, London.

Bresnan, M. (1996), 'An organizational perspective on changing buyer-supplier relations: a critical review of evidence', *Organization*, vol. 3, no. 1, pp. 121-146.

Brusco, S. (1996), 'Trust, social capital and local development: some lessons from the experience of the Italian districts', in OECD (ed), *Networks of Enterprises and Local Development: Competing and Co-operating in Local Productive Systems*, LEED for OECD, Paris, pp.115-119.

Callon, M. (1986), 'Some elements of a sociology of translation: domestication of the scallops and the fishermen of St Brieue Bay', in J. Law (ed), *Power Action and Belief: A New Sociology of Knowledge*, Routledge, London, pp. 196-233.

Clark, G. (1997), 'Rogues and regulation in global finance: Maxwell, Leeson and the City of London', *Regional Studies*, vol. 31, pp. 221-236.

Clegg, S. (1989), *Frameworks of Power*, Sage, London.

Clegg, S. (1990), *Modern Organizations*, Sage, London.

Conti, S., Malecki, E. and Oinas, P. (eds), *The Industrial Enterprise and Its Environment: Spatial Perspectives*, Avebury, Aldershot.

Cooke, P. (1998), 'Introduction: origins of the concept', in H-J. Braczyk, P. Cooke and M. Heidenreich (eds), *Regional Innovation Systems*, UCL Press, London and Bristol PA, pp. 2-25.

Crew, L. (1996), 'Material culture: embedded firms, organizational networks and the local economic development of a fashion quarter', *Regional Studies*, vol. 30, pp. 257-272.

Department of Trade and Industry (2001), *Business Clusters in the UK – A First Assessment*, volumes 1, 2 and 3, 238pp.

Dicken, P. (1994), 'Global-local tensions: firms and states in the global space economy', *Economic Geography*, vol. 70, no. 1, pp. 101-128.

Dicken, P., Forsgren, M. and Malmberg, A. (1994), 'The local embeddedness of transnational corporations', in A. Amin, and N. Thrift (eds), *Globalization, Institutions and Regional Development in Europe*, Oxford University Press, Oxford, pp. 21-25.

Dicken, P. and Thrift, N. (1992), 'The organization of production and the production of organization; why business enterprises matter in the study of geographical industrialization', *Transactions of the Institute of British Geographers, New Series*, vol. 17, pp. 279-291.

Dunning, J. (1979), 'Explaining changing patterns of international production: in defence of the eclectic theory', *Oxford Bulletin of Economics and Statistics*, vol. 41, pp. 249-265.

Enright, M. (1995), 'Regional clusters and economic development: a research agenda', U. Staber, B. Schaefer and B. Sharma (eds), *Business Networks: Prospects for Regional Development*, de Gruyter, Berlin, pp. 190-214.

Forbes, D. and Rimmer, P. (eds) (1984), *The Geographical Transfer of Value*, Monograph HG17, Department of Human Geography, Australian National University, Canberra.

Fröbel, F., Heinrichs, J. and Kreye, O. (1980), *The New International Division of Labor*, Cambridge University Press, Cambridge.

Fujita, K. and Hill, R. (1995), 'Global toyotaism and local development', *International Journal of Urban and Regional Research*, vol. 19, pp. 7-22.

Gereffi, G. (1994), 'The organization of buyer-driven global commodity chains: How US retailers shape overseas production networks', in G. Gereffi and M. Korzeniewicz (eds), *Commodity Chains and Global Capitalism*, Praeger, Westport CT, pp. 67-92.

Gereffi, G. (1999), 'International trade and industrial upgrading in the apparel commodity chain', *Journal of International Economics*, vol. 48, pp. 37-70.

Glasmeier, A. (1991), 'Technological discontinuities and flexible production networks: The case of Switzerland and the world watch industry', *Research Policy*, vol. 20, pp. 469-485.

Grabher, G. (1993a), 'Rediscovering the social in the economics of interfirm relations', in G. Grabher (ed), *The Embedded Firm: On the Socioeconomics of Industrial Networks*, Routledge, London and New York, pp. 1-31.

Grabher, G. (ed) (1993b), *The Embedded Firm: On the Socioeconomics of Industrial Networks*, Routledge, London.

Granovetter, M. (1985), 'Economic action and social structure: The problem of embeddedness', *American Journal of Sociology*, vol. 91, no. 3, pp. 481-510.

Hallsworth, A. and Taylor, M. (1996), "Buying Power': Interpreting retail change in a circuits of power framework', *Environment and Planning A*, vol. 28, pp. 2125-2137.

Hambrick, D. and Mason, P. (1984), 'Upper echelons: the organization as a reflection of its top managers', *Academy of Management Review*, vol. 9, pp. 193-206.

Heidenreich, M. and Krauss, G. (1998), 'The Baden-Württemberg production and innovation regime: Past success and new challenges', in H-J. Braczyk, P. Cooke and M. Heidenreich (eds), *Regional Innovation Systems: The Role of Governance in a Globalized World*, UCL Press, London, pp. 214-244.

Honig, B. (1998), 'Who gets the goodies? An examination of microenterprise credit in Jamaica', *Entrepreneurship and Regional Development*, vol. 10, no. 4, pp. 313-334.

Hsing, Y-T. (1999), 'Trading companies in Taiwan's fashion shoe networks', *Journal of International Economics*, vol. 48, pp. 101-120.

Hudson, R. (1999), 'The learning economy, the learning firm and the learning region: A sympathetic critique of the limits of learning', *European Urban and Regional Studies*, vol. 6, no. 1, pp. 59-72.

Jonas, A. (1996), 'Local labour control regimes: uneven development and the social regulation of production', *Regional Studies*, vol 30, no. 4, pp. 323-338.

Korzeniewicz, M. (1992), 'Global commodity networks and the leather industry: Emerging forms of economic organization in a postmodern world', *Sociological Perspectives*, vol. 35, pp. 313-327.

Leborgne, D. and Lipietz, A. (1992), 'Conceptual fallacies and open questions on post-fordism', in M. Storper and A. Scott (eds), *Pathways to Industrialization and Regional Development*, Routledge, London, pp. 332-348.

Lovering, J. (1999), 'Theory led by policy: the inadequacies of "the new regionalism"', *International Journal of Urban and Regional Research*, vol. 23, no. 2, pp. 379-395.

Lundvall, B-Å. (1993), 'Explaining interfirm cooperation and innovation: Limits of the transaction cost approach', in G. Grabher (ed), *The Embedded Firm: On the Socioeconomics of Industrial Networks*, Routledge, London and New York, pp. 52-64.

Maskell, P. (1999), 'Social capital, innovation and competitiveness', unpublished manuscript.

Maskell, P., Eskilinen, H., Hannibalsson, I., Malmberg, A. and Vatne, E. (1998), *Competitiveness, Localized Learning and Regional Development – Specialization and Prosperity in Small Open Economies*, Routledge, London.

Maskell, P. and Malmberg, A. (1999), 'Localised learning and industrial competitiveness', *Cambridge Journal of Economics*, vol. 23, pp. 167-190.

Massey, D. (1984), *Spatial Divisions of Labour*, Macmillan, London.

McNulty, T. and Pettrigrew, A. (1999), 'Strategists on the board', *Organization Studies*, vol. 20, no. 1, pp. 47-74.

Mohan, G. (1996), 'Neoliberalism and decentralised planning development', *Third World Planning Review*, vol. 18, no. 4, pp. 433-454.

Ohmae, K. (1995), *The End of the Nation State: The Rise of Regional Economies*, Harper Collins, London.

Oinas, P. (1997), 'On the socio-spatial embeddedness of business firms', *Erdkunde*, vol. 51, pp. 23-32.

Pfeffer, J. (1981), *Power in Organizations*, Pitman, Boston MA.

Piore, M.J. and Sabel, C.F. (1984), *The Second Industrial Divide: Possibilities for Prosperity*, Basic Books, New York.

Polanyi, K. (1957), 'The economy of instituted process', in K. Polanyi, C. Arensberg and H. Pearson (eds), *Trade and Markets in Early Empires*, Free Press, Glencoe IL, pp. 243-270.

Porter, M.E. (1998), *On Competition*, Macmillan, London.

Powell, W.W. and DiMaggio, P.J. (eds) (1991), *The New Institutionalism in Organizational Analysis*, The University of Chicago Press, Chicago IL.

Powell, W.W. and Smith Doerr, L. (1994), 'Networks and economic life', in N. Smelser and R. Swedberg (eds), *The Handbook of Economic Sociology*, Princeton UP, Princeton NJ, pp. 368-402.

Pratt, A. (1997), ' The emerging shape and form of innovation networks and institutions', in J. Simmie (ed), *Innovation, Networks and Learning Regions?*, Jessica Kingsley Publishers, London and Bristol PA and Regional Studies Association, London, pp. 124-136.

Pryke, M. and Lee, R. (1995), 'Place your bets: towards an understanding of globalisation, socio-financial engineering and competition within a financial centre', *Urban Studies*, vol. 32, pp. 329-344.

Putnam, R.D. (1993), *Making Democracy Work: Civic Traditions in Modern Italy*, Princeton University Press, Princeton.

Rasmussen, J., Schmitz, H. and van Dijk, M. (1992), 'Introduction: Exploring a new approach to small-scale industry', *IDS Bulletin*, vol. 23, no. 3, pp. 2-6.

Sabel, C. (1989), 'Flexible specialization and the re-emergence of regional economies', in P. Hirst and J. Zeitlin (eds), *Reversing Industrial Decline? Industrial Structure and Policy in Britain and Her Competitors,* Berg, Oxford, pp. 17-70.

Schmitz, H. (1993), 'Small shoemakers and fordist giants: tales of a supercluster', *IDS Discussion Paper No. 331*, IDS, Sussex.

Schmitz, H. and Musyck, B. (1993), 'Industrial Districts in Europe: Policy Lessons for Developing Countries?', *IDS Discussion Paper No. 324*, IDS, Sussex.

Simmie, J. (ed), (1997), *Innovation, Networks and Learning Regions?* Regional Policy and Development Series, Jessica Kingsley Publishers, London and Bristol PA and Regional Studies Association, London.

Scott, A.J. and Storper, M. (1992), 'Industrialization and regional development', in M. Storper and A. Scott (eds), *Pathways to Industrialization and Regional Development*, Routledge, London, pp. 3-17.

Staber, U. (1996), 'Accounting for differences in the performance of industrial districts', *International Journal of Urban and Regional Research*, vol. 20, no. 2, pp. 299-316.

Storper, M. (1997), *The Regional World: Territorial Development in a Global Economy*, Guilford Press, New York.

Taylor, M. (1995), 'The business enterprise, power and patterns of geographical industrialization', in S. Conti, E. Malecki and P. Oinas (eds), *The Industrial Organisation and Its Environment: Spatial Perspectives*, Avebury, Aldershot, pp. 99-122.

Taylor, M. (2000), 'Enterprise, power and embeddedness: an empirical exploration', in E. Vatne and M. Taylor (eds), *The Networked Firm in a Global World*, Ashgate, Aldershot, pp. 199-234.

Taylor, M. and Conti, S. (eds) (1997), *Interdependent and Uneven Development: Global-Local Perspectives*, Ashgate, Aldershot.

Taylor, M., Ekinsmyth, C. and Leonard, S. (1997), 'Global-local interdependencies and conflicting spatialities: "space" and "place" in economic geography', in M. Taylor and S. Conti (eds), *Interdependent and Uneven Development: Global-Local Perspectives*, Ashgate, Aldershot, pp. 57-80.

The Economist (1995), *A Survey of Multinationals: Big Is Back*, vol. 335, No. 7920, June, pp. 24-30.

Thrift, N. (1998), 'The rise of soft capitalism', in A. Herod, G. ÓTuathail, and S. Roberts (eds), *An Unruly World? Globalization, Governance and Geography*, Routledge, London and New York, pp. 25-71.

Uzzi, B. (1996), 'The sources and consequences of embeddedness for the economic performance of organizations: The network effect', *American Sociological Review*, vol. 61, pp. 674-698.

Uzzi, B. (1997), 'Social structure and competition in interfirm networks: The paradox of embeddedness', *Administrative Science Quarterly*, vol. 42, no. 1, pp. 35-67.

Wrong, D.H. (1995), *Power: Its Forms, Bases and Uses*, Transaction Publishers, New Brunswick and London.

Yeung, H. (1997a), 'Cooperative strategies and Chinese business networks: a study of Hong Kong transnational corporations in the ASEAN region', in P. Beamish and J. Killing (eds), *Cooperative Strategies: Asia-Pacific Perspectives*, The New Lexington Press, San Francisco CA.

Yeung, H. (1997b), 'Business networks and transnational corporations: a study of Hong Kong firms in the ASEAN region', *Economic Geography*, vol. 73, no. 1, pp. 1-25.

Yeung, H. (1998a), 'Capital, state and space: contesting the borderless world', *Transactions of the Institute of British Geographers, New Series*, vol. 23, pp. 291-309.

Yeung, H. (1998b), 'The social-spatial constitution of business organizations: a geographical perspective', *Organizations*, vol. 5, no. 1, pp. 101-128.

Zukin, S. and Di Maggio, P. (eds) (1990), *The Social Organization of the Economy*, Cambridge University Press, Cambridge.

Chapter 2

Embeddedness and Innovation

Ron Boschma, Jan Lambooy and Veronique Schutjens

Introduction

There is a growing body of literature on 'embeddedness'. The idea had been introduced by Polanyi (1944) and Granovetter (1985) to indicate that economic relations co-exist with a set of social attributes. Economic actors make their decisions not only on information about prices and quantities, but also on other attributes, like trust and social values. Market relations in economic theories are often devoid of 'humanness'. In traditional economic textbooks, firms are almost always seen as closed systems of production (the black box of the production function) without relevant relations other than those via the market, and only reacting to price and quantity signals. Optimal inputs of resources are evaluated only on the configuration of prices, not on social values. This contrasts with the 'openness' and the 'fuzzy boundaries' of firms in a broader perspective where they are regarded as an integrated part of complex social structures and relations.

Economic geographers have been eager to embrace the notion of embeddedness because it assumes that firms are closely linked to their local production environments in a world of increasing globalisation. 'Embeddedness' not only accounts for the importance of trust-based networks for regional development, it also incorporates the idea that socio-cultural and institutional factors (as well as 'traded interdependencies') may be essential (Storper 1997).

However, there is an urgent need to unravel the notion of embeddedness. According to Oinas (1997), 'the notion of embeddedness seems to capture all possible aspects in a firm's environment. This is why it is problematic: it encompasses too many things with the result of being ambiguous' (p. 26). Taylor and Leonard (this volume) state that the concept of embeddedness is imprecise and ill defined. The same may be said about its application in economic geography. Although widely used among economic geographers, Oinas (1998) remarks that, '... it has remained far from clear what embeddedness actually refers to' (p. 49).

This chapter has two objectives. The first objective is to clarify the different interpretations of what embeddedness means. The aim is to shed more light on the analytical content of the notion; a research topic that has hardly been covered. We distinguish between two levels at which social relations may affect economic actions and outcomes. On the one hand, there are the 'social relations with known actors' that involve both dyadic relations and network relations, which are associated with the social embedding of inter-firm relationships. On the other hand, there is the broader, contextual meaning of embeddedness, which we associate with, among other things, the socio-cultural and institutional environment firms operate in. As part of this objective we explore the impact this contextual dimension of embeddedness might have on the nature of firms' buyer-supplier relationships. In this respect, we also introduce a spatial dimension because, among other reasons, spatial units (on whatever level) may provide a socio-cultural environment that is more likely to generate embedded inter-firm relationships.

The second objective of the chapter is to shed light on the (economic and innovative) performance of embedded firms. As Uzzi (1996) noted, 'the concept of embeddedness does not explain concretely how social ties affect economic outcomes' (p. 674). Embeddedness as such does not clarify the processes of adjustment of entrepreneurs. It only points to the fact that economic actors use more variables than prices and profits alone on which to base their decisions. Moreover, it is far from clear whether embeddedness may have the positive economic effects suggested by Granovetter's notion of the 'strength of weak ties'. In fact, it may have adverse impacts because of 'lock-in' effects. Based on this debate, we propose an inverted-U relationship between the degree of embeddedness and the innovative performance of a firm.

Following the introduction, the chapter is divided into four sections. In Section 2, we explore the concept of embeddedness, its theoretical background, and the different interpretations it has been given. In Section 3, we introduce the micro-level of embeddedness by indicating what are the main features of embedded inter-firm relationships. In Section 4, we examine the macro-level of embeddedness, in an attempt to specify the impact of the socio-cultural environment (defined as social capital) on the nature of inter-firm relationships sketched out in the previous section.

In Section 5, we return to the micro-level of embeddedness. We focus on the effects of embedded inter-firm relationships on the innovative performance of firms and the debate that surrounds this issue. From a neo-classical economics perspective it can be claimed that embedded relationships are inefficient and have a negative impact on the performance of firms. However, the embeddedness literature suggests the opposite: the more embedded the relationships of a firm, the better its performance, because the firm's social relationships encourage the exchange of (tacit) knowledge, and thus facilitate learning and innovation (Boschma 1999b). Following Uzzi (1997), we go beyond this debate by proposing an inverted-U relationship between embeddedness and innovative performance at the firm level. We suggest that embeddedness has a positive influence on the performance of a firm up to a certain threshold, after which point adverse impacts arise because of lock-in. In Section 6, we draw together the main conclusions.

The Notion of Embeddedness

Following Granovetter (1985), the embeddedness literature has blamed neo-classical economics (including Williamson's 'new institutional economics') for an 'under-socialised' view of economic relations, that emphasises rational, self-interested behaviour that is hardly affected by social relations. Neo-classical economics regards actors as 'atomistic' individuals who act independently and maximise their utility, and exchange goods and services in one-off deals, based solely on price and quality signals. Their actions are devoid of social context – divorced from social norms, social networks and trust (Coleman 1990). Analogous with Adam Smith's ideas, this view gives a normative basis to the free market, in which social relations are regarded as obstacles to the proper functioning of competitive markets.

In contrast, Polanyi (1944) and Granovetter (1985) proposed the notion of embeddedness to indicate that economic relations are embedded in social relations that go beyond strictly rational or monetary values. Economic relations also encompass other aspects, '... on the basis of trust and its sources: ethics, kinship, friendship or empathy' (Nooteboom 1999, p. 24). This embeddedness perspective sees personal ties of trust and loyalty (rather than impersonal transactions) and stable relations between exchange partners (rather than one-off exchanges) as elements that support economic behaviour and development (Håkansson and Johanson 1993).

The interactive form of economic coordination is likely to generate trust and to discourage opportunistic behaviour (Fukuyama 1995). The potential of trust-based relations to solve opportunistic behaviour has largely been ignored by Williamson (Oerlemans 1996). As Harrison (1992) puts it,

> ... the institutional mechanisms that preoccupy the Williamsonians – elaborate explicit and implicit contracts – are actually a functional *substitute* for trust, not a set of social arrangements or ongoing relationships that might cultivate that sensibility. Where formal incentives and sanctions are the methods for reducing shirking on the job or the violation of 'intellectual property rights' (company secrets), there *is* no active role for social relations; these become mere window dressing. (p. 477)

The notion of embeddedness has also changed the characterisation of the firm itself. Firms are regarded as 'open systems'. This creates a certain 'fuzziness' in the boundaries of production units and firms. As Harrison (1992) puts it, 'firms relate to one another by interpenetrating one another's formal organizational boundaries, rather than solely through the price-mediated exchange of commodities' (p. 478). This implies that the acquisition of some inputs, like knowledge, is not a clear matter either. Knowledge as a resource is difficult to contain within one production unit or firm for a long time, because it is related to human experience and interaction, and thus a collective process, sensitive to social values (Lambooy 1997).

In economic geography, Granovetter's notion of embeddedness has been widely adopted because it revives the idea that firms are firmly linked to their local production environment in a world of increasing globalisation. Harrison (1992), for example, has applied this notion to the phenomenon of 'industrial districts': 'in the concept of embeddedness lies the key to understanding how the theory of industrial districts differs fundamentally from neoclassical agglomeration theory' (p. 476). According to Harrison (1992), the industrial district is all about 'interdependence of firms, flexible firm boundaries, co-operative competition and the importance of trust in reproducing sustained collaboration among economic actors ...'(p. 471). Storper (1997) argues that not only market relations are important, but that more encompassing systems of interdependencies need to evolve, especially within regions: 'untraded interdependencies'. These are based on experience, trust and convention. One conclusion of Storper's argument is that enterprises cannot really survive without non-market interdependencies, which are a condition for and a result of a certain embeddedness of economic actors in regional structures.

Having said this, we feel there is an urgent need to unravel this notion of embeddedness. Even Granovetter (1985) warned that the concept can easily become an empty box that explains anything. After a review of the literature, Oinas (1998) concluded that, '... just about any interpretation seems to be legitimate. This renders the term next to meaningless' (p. 53).

Therefore, we distinguish two levels at which social relations may affect economic behaviour and performance. At the first level, there are the social relations with known actors that involve dyadic relations or a larger network of relations (mostly based on bonds of friendship or kinship) that we associate with the social embedding of inter-firm relationships. This interpretation refers to the social dimension of economic relationships between firms at the micro-level, which is explained in more detail in the next section.

However, embeddedness at the micro-level is a rather static concept. It only points to the fact that economic actors use more variables than purely economic ones (like prices and profits) to make their strategic decisions. This interpretation leaves no room for changing attitudes and changing economic relationships. Therefore, we also use a second, broader, macro contextual meaning of embeddedness, which we associate in Section 4 with the socio-cultural context that firms operate within. The contextual dimension is closely related to institutional economics, which focuses not so much on markets as such, but rather on the transactions among the participants in those markets, bound by the values and rules of the relevant society. As a result, this social context may explain the degree of embeddedness of inter-firm relationships at the micro-level.

Between these macro- and micro-level perspectives, it is also possible to distinguish a meso-level. This meso-level dimension focuses on the embeddedness of firms within a network (Uzzi 1996, Nooteboom 1999). Although we use some concepts from network analysis to explain the characteristics of inter-firm relationships in more detail, for the purpose of our argument, we concentrate only on the macro-level and micro-level dimensions of embeddedness.

The Micro-level Dimension of Embeddedness

As explained in the previous section, the more limited meaning of embeddedness is associated with the social embedding of economic relations. Here, we elaborate on insights of 'transaction costs economics' and 'network analysis' in order to build a picture of the main differences between market-based and embedded economic relationships. This implies that embeddedness may be of different degrees, and that the neo-classical form of economic transaction should not be ruled out just because it ignores the social dimension. It is very likely that reality involves a mix of both types of transaction creating a continuum of embeddedness (Uzzi 1997).

In Table 2.1, we give a short overview of what we believe are the key contrasts between market-based relationships and embedded relationships at the micro-level. First, the embedded form of economic exchange involves a sequence of reciprocated transactions. As it was put by Uzzi (1996), 'network theory argues that embeddedness shifts actors' motivations away from the narrow pursuit of immediate economic gains towards the enrichment of relationships through trust and reciprocity' (p. 677). Second, profit-maximising motives guide the market-based transaction that is limited to the exchange of information on price and quality. This stands in contrast to embedded relationships: 'as Lundvall (1993) demonstrates, interactive learning presupposes an orientation to "communicative rationality"- that is, an orientation to understanding which transcends the narrow market calculus of minimizing transaction costs' (Grabher 1993, p. 10). Third, the stability of network-based interactions leads to 'interdependence' between actors, rather than the 'discrete' exchange relations between 'independent' actors in markets, and coordination processes within hierarchies (implying 'dependence'). Fourth, conflicts in embedded ties are resolved through voice-based strategies based on 'give and take' (and joint-problem solving) rather than 'exit' strategies (walking out) in market exchanges. Finally, embedded relationships are characterised by 'loose coupling', in which the exchange partners retain some autonomy, while stimulating the exchange of 'tacit' knowledge.

Grabher (1993) maintains that networks consist of actors possessing power, which he regards as a functional element of networks. We do not deny the existence of power in economic relationships (see Taylor and Leonard, this volume), but we believe that power is an element that obscures rather than clarifies the lines drawn between market-based and embedded relationships. In market-based relationships, power may, or may not, be involved, depending on the positions of trading actors. The same applies for embedded relationships, which is reflected by the different positions in the literature with respect to the impact of power in networks on innovative performance. Broadly speaking, the 'industrial district' literature puts emphasis on the collaboration between equal network partners, which is regarded as a stimulus for interactive learning. In contrast, Grabher (1993) points out that radical change requires power in networks. Moreover, Taylor (1999) has stressed that network partners often follow strategies of exclusion, which are to the detriment of newcomers and new initiatives.

Table 2.1   Main characteristics of market-based relationships and embedded relationships

Market-based relationships	*Embedded relationships*
Arm's-length transactions	Network form of economic exchange
Narrow pursuit of immediate economic gains by self-interested actors	Relationship through trust and reciprocity
Profit-maximising rationality	Communicative rationality
Independence: discrete exchange relations	Interdependence: concentrated exchange relations
'Exit-based' strategy to solve problems (competition)	'Voice-based' strategy to solve problems (give and take)
No coupling	Loose coupling

The Macro-level Dimension of Embeddedness

As explained above, next to the micro-level we also distinguish a contextual dimension of embeddedness at the macro-level, which we associate with the socio-cultural environment firms operate in. North (1990) has made a distinction between the 'institutional environment' at the macro-level (such as norms and values of conduct) and 'institutional arrangements' at the micro-level, in which these norms and values are embodied in specific exchange relations. We link the contextual dimension of embeddedness at the macro-level to the relational dimension of embeddedness at the micro-level, *in casu* the nature of supplier relationships, which is an issue Uzzi has largely ignored. By doing so, we overcome the 'undersocialised' view of mainstream economics when keeping in mind that economic actors operate in a social context.

According to Zukin and Di Maggio (1990), the contextual dimension of embeddedness consists of many aspects. *Political embeddedness* identifies economic actors embedded in the institutional 'rules of the game' (for instance, political decision making structures, laws and fiscal rules). *Cultural embeddedness* refers to sets of shared values, like trust, that may be specific to a group of entrepreneurs (e.g. ethnic and religious values). Finally, *cognitive embeddedness* deals with the knowledge aspects of embeddedness. For reasons of simplicity, we restrict our attention to the notion of 'social capital', which comes close to the broad notion of cultural embeddedness.

Social capital covers more or less the broad, contextual meaning of embeddedness: the socio-cultural context. Social capital is an idea that has attracted a great deal of attention in economics (Fukuyama 1995, Bolton 1998). Coleman (1990), among others, has stated that it is very difficult to grasp the essence of social capital because of its intangible nature. According to Morgan (1997), 'social capital refers to features of social organisation, such as networks, norms and trust, that facilitate co-ordination and co-operation for mutual benefit' (p. 493). In this way, social capital is seen as a productive resource within a structure of relations available to actors (Coleman 1990).

The notion of social capital is considered relevant here, because it puts emphasis on the nature of socio-cultural relations that may (or may not) bring and hold together (economic) actors. In this respect, it functions as a sort of 'glue' for collective action. Therefore, this concept may bridge the gap between the contextual dimension of embeddedness at the macro-level and the social embedding of inter-firm relationships at the micro-level. More precisely, social capital may constitute a productive resource that enables cooperation and increases performances of firms due to lower transaction costs.

Social capital has been seen as the cause of the post-war growth of many industrial districts in the Third Italy. A long-term empirical analysis has demonstrated that the distinctive social structure of the Third Italy (in terms of common values and norms) provided a basis on which these local networks of SMEs could emerge and flourish (Boschma 1999a). A homogeneous and shared culture is more likely to generate tight inter-firm relationships (Saxenian 1994). In this respect, cultural proximity, among other factors, brought about external scale economies, lower transaction costs, collective learning processes and flexibility.

Moreover, geographical proximity can stimulate embedded inter-firm relationships. Short distances favour information contacts and exchange among actors (Schutjens and Stam 2000). Geographical proximity also facilitates informal relationships (Audretsch and Stephan 1996). Firms located near to each other have more face-to-face contacts and can easily build up trust, which leads to more personal and thus embedded relationships between firms. According to Harrison (1992), this logic runs from, '... proximity to experience to trust to collaboration to enhanced regional economic growth' (p. 478). Grabher (1993) argues that cultural and geographical proximity are intertwined and reinforce each other: '... the homogeneous culture creates rules and engenders trust, and its geographic boundedness increases the probabilities of social interaction and communication that reduce the problem of bounded rationality' (p. 22).

This is not to say that embedded relations are necessarily of a local nature. Although often suggested otherwise, embeddedness may well have a non-local dimension (Oinas 1997). In this respect, Hausmann's (1996) distinction between organisational, social and spatial proximity is relevant. He claims that for inter-organisational learning, social or organisational proximity may be more important than spatial proximity, but that spatial proximity strongly facilitates these effects.

The Impacts of Embeddedness on the Economic Performance of Firms

In this section, we return to the micro-level of embeddedness. As noted before, the concept of embeddedness itself does not provide an explanation for economic performance. We describe what impact embeddedness might have on economic performance at the firm level. There are many aspects to this question, but here we focus on the micro-level of inter-firm relationships. Moreover, since economic performance covers so many areas (e.g. profit, productivity, turnover, shareholder's value, and so on), we confine ourselves to innovative performance. This allows us to link more directly the notion of embeddedness to the adaptive capability of firms, which is essential for their competitiveness. We outline the positions adopted in the economics literature, which reduce to the positive aspects ('strength of weak ties') and negative sides ('lock-in') of embedded inter-firm relationships. We synthesize the results of this debate, and propose an inverted-U relationship between the degree of embeddedness of a firm and its innovative performance.

Granovetter's principle of 'strength of weak ties' underpins the positive effects of embeddedness on micro-level economic performance. Four advantages of embeddedness can be recognised: lower transaction costs (mainly contracting and monitoring costs); new cooperative institutions; greater flexibility; and organisational learning. Closely related to these factors are the characteristics of embedded inter-firm relationships shown in Table 2.1.

First, trust-based supplier relationships reduce (but do not eliminate) the risk of opportunistic behaviour by exchange partners. This reduces transaction costs on the specification and monitoring of contracts. When trust is high, there is less need to specify all the details of a transaction in formal written contracts. Norms that are shared effectively constrain opportunistic behaviour. The need to control and monitor transactions is also reduced. Thus, transactions based on trust and shared norms are more efficient, less costly and more effective (Putnam 1993).

Second, cooperation is vital for the competitiveness of the small firms. Small firms tend to lack the resources to be successful in export markets, to do their own research, to negotiate with large banks for loans on favourable terms, and so on. Trust in relationships can bring these advantages (Dei Ottati 1994). As Harrison (1992) states, 'firms are said to co-operate on getting new work into the district, in forming consortia to obtain cheap credit, in jointly purchasing raw materials, in bidding on large projects and in conducting joint research' (p. 478).

Third, embedded relationships (or 'loose couplings') greatly enhance flexibility, because the partners retain some autonomy, and autonomy prevents lock-in. In strongly embedded networks, independent and autonomous firms both fiercely compete and closely cooperate. With good communications, independent partners may be able to shift goals and strategies more easily. This is different for market-based transactions. According to Nooteboom (1999):

> with detailed formal contracting, it is more difficult (slow and costly) to modify terms when conditions change. It yields a straightjacket for action which can be very constraining especially when the goal of the relation is innovation: the development or implementation of novelty. Then virtually by

definition one cannot foresee what duties are to be regulated and what returns are to be shared. (p. 25)

Fourth, embedded relationships favour the transmission and exchange of knowledge and information and, thus, learning and innovation (Boschma 1999b). Because of joint-problem solving, trust and dense information exchange, embedded relationships stimulate interactive learning and thus, innovations (Asheim 1996). They not only lower the costs of search, trust-based relationships also facilitate the exchange of tacit knowledge which is, by nature, much more difficult to communicate and to trade through markets (Maskell and Malmberg 1999). Embedded relationships reflect a social and open attitude of 'communicative rationality' (rather than a pure, calculative and narrow market orientation towards minimising costs), which is conducive to (and even a prerequisite for) interactive learning (Lundvall 1993). Moreover, interactive learning requires durable, committed and long-term relationships.

The negative effects of embeddedness on micro-level economic performance can be associated with the 'weakness of strong ties', or 'lock-in'. In essence, this comes down to a poor ability to interpret new information or an incapability to adjust accordingly. There may be several reasons for this, such as loyalty (especially when friendship or kinship is involved), long-term commitment and cognitive lock-in (in the sense that routines in inter-firm relationships obscure the view on new technologies or new market possibilities). These arguments are briefly explained below.

First, embedded ties hold the possibility of underestimating opportunism, especially when relations are based on emotional bonds of friendship and kinship. According to Nooteboom (1999), '... strong ties may have the disadvantage ... of generating too much personal interaction and loyalty, to the detriment of productive work, criticism and flexibility' (p. 13). In this respect, embedded relationships, based on trust and positive values, provide firms with a false sense of certainty. Accordingly, too much 'social behaviour' may have negative consequences in a world with calculating actors, in markets where technologies and policies continually change in conditions of uncertainty, and where opportunism is a common attitude. As Uzzi (1996) has put it, 'in highly embedded networks, feelings of obligation, friendship or betrayal may ... be so intense that emotions override economic imperatives' (p. 684).

Second, long-term or too much commitment may lock buyers and suppliers into established ways of doing things or into specific technological trajectories, at the expense of their own innovative and learning capacity. In this way, they not only become sealed off from new market developments. Even if they have wide access to new information, they are incapable of adapting. In evolutionary thinking, this has been explained in terms of 'routines', 'path-dependent' behaviour, and cognitive 'lock-in' (Boschma and Lambooy 1999, Hudson 1999). Switching costs are especially high in the case of radical change (Shapiro and Varian 1999). Moreover, there may also be huge costs involved in building up trust between partners (Gambetta 1988).

In sum, there is considerable controversy in the economics literature about the effects of embeddedness on micro-level economic performance. On the one hand, neo-classical economics would claim that embedded relationships are inefficient and will have a negative impact on the performance of firms. On the other hand, the embeddedness literature would say the opposite: the more embedded the relationships of a firm are, the more superior its performance because its social relationships facilitate the communication and exchange of (tacit) knowledge, and thus learning and innovation.

Following Uzzi (1996, 1997), we make an attempt to go beyond this debate by integrating these two contrasting views into an inverted-U shaped relationship between embeddedness and innovative performance at the firm level. We propose that the social dimension of economic relationships has a positive influence on the performance of a firm up to a certain threshold (contrary to neo-classical thinking), after which adverse impacts come into being as a result of lock-in (contrary to the embeddedness model).

This idea is depicted in Figure 2.1 as the Uzzi model. At first, embedded inter-firm exchange promotes the economic performance of firms. Collective agreements, shared investments, concentrated exchange of information and knowledge between partners, characterised by interdependency, reciprocity and trust, lower transaction costs, reduce risks and uncertainty, and increase the access to tacit knowledge. In other words, social relationships stimulate learning and innovation. However, these positive effects can turn negative when the embedded relationships become too closely tied. Then, the economic efficiency becomes vulnerable to unforeseen, exogenous shocks that may ruin previously safe exchange relationships, especially when these relationships are rather isolated from vital external information on markets or technology (Camagni 1991). At the micro-level of the firm and the meso-level of the network this equates with insufficient information exchange with the world outside, resulting in a loss of competitiveness.

This expected relationship has been to some extent observed and verified in various empirical studies by Uzzi at the meso-level of the network (1996, 1997, 1999). Uzzi's empirical analysis of the New York apparel industry led to the conclusion that 'firms organized in networks have higher survival chances than do firms which maintain arm's-length market relationships. The positive effect of embeddedness reaches a threshold, however, after which point the positive effect reverses itself' (Uzzi 1996, p. 674). Uzzi suggests that an optimum in terms of adaptive capacity can be reached when the network of a firm consists of a mixture of arm's-length ties and embedded relationships.

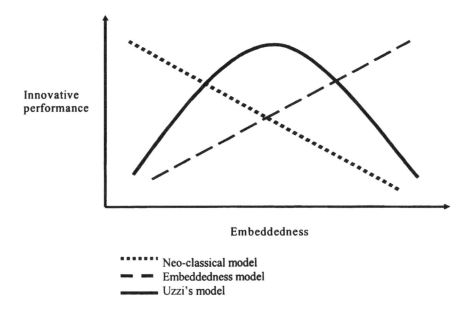

Figure 2.1 The relationship between the degree of embeddedness and the innovative performance of a firm

Conclusions

Our main objective was to clarify the meaning and relevance of the concept of embeddedness. This is a challenging task because this notion has something promising to offer. We firmly believe that the behaviour of economic actors may be anchored in trust-based inter-firm networks in which personal relations, long-term interdependence, knowledge exchange and reciprocity are key elements. In this respect, embeddedness may be of great importance for the competitiveness of firms and, in the end, the economic performance of regions.

Much of the literature on embeddedness lacks clarity owing to the many different interpretations the term 'embeddedness' has been given. We distinguished between a micro-level dimension of embeddedness (i.e. the social embedding of economic relationships) and a macro-level dimension of embeddedness. This allowed us to link the contextual dimension of embeddedness (here defined as social capital) to the nature and the quality of exchange relationships between firms. Still, we think this concept should be explored more thoroughly, not only conceptually, but also, and even more so, empirically. Many analytical problems remain before the meaning and impact of embeddedness can be adequately demonstrated. As a conclusion to the discussion of the chapter, we briefly elaborate some of the more important of these research challenges.

First, embeddedness is hard to measure empirically. Partly this is because embeddedness may appear at various levels. Generally, the higher the level of analysis, the harder it is to define or measure embeddedness. How is it possible to prove, for example, that a firm is embedded in its local environment? According to Oinas (1997), the embeddedness of a firm should be interpreted somehow as the sum of individual embeddedness of key decision-makers (owners, managers, key employees) in a firm, and the embeddedness of the firm as a collective actor in its external environment. There are similar measurement problems at the macro-level dimension of embeddedness. Although we believe embedded inter-firm relationships are more likely to develop in environments endowed with social capital, there is hardly any empirical evidence for this, because of the fact that the stock of social capital is so hard to measure (Boschma 1999a).

Second, it remains difficult to operationalise the relation between the social context at the macro-level and embedded relationships at the micro-level. This is because firm strategies may actively shape and change their social relations and their surrounding socio-cultural environment. We think the interdependence of both levels of embeddedness is a very important issue, which is often ignored.

Third, we urgently need in-depth analyses at the micro-level that develop explanations of embedded relationships. Apart from the impact of social capital discussed in this chapter, there is as yet little known about what other factors may be involved. For example, there may be a positive relationship between the degree of complexity and the dynamics of the local environment and the need for embeddedness. However, the opposite may also be true. The more complex and the more dynamic societies are, the more likely it is that institutions have developed that handle transactions and uncertainty (such as a formal banking system and a legal system that enforce contracts). Consequently, the less need there may be for trust-based relationships. Moreover, it seems important to control for the size of firms and the durability of relationships, although it is still uncertain what impact these parameters might have on the social embedding of economic relations. We might, for instance, argue that small firms rely on more informal relationships because they face many restrictions in terms of the contact, contract and control dimensions of their transactions (Nooteboom 1999). Another argument is that, the more long-standing relationships are, the more trust will be built into them and, therefore, the more embedded economic relationships become (Schutjens and Stam 2000).

Fourth, there is considerable controversy in the literature about the relationship between embeddedness and innovative performance. Our analysis highlighted issues concerning the direction of the embeddedness effect, particularly whether socially embedded inter-firm relationships will really stimulate innovative performance, and if so, up to what point. We attempted to visualise the anticipated combination of positive and negative effects of embedded inter-firm relationships on innovative performance as an inverted U-shaped relationship by extending Uzzi's work on networks. However, there is a need to identify what factors shape this inverted U. relationship. We need to conceptualise how different socio-cultural environments might affect the slope, width and height of this relationship. Just as importantly, we need sound empirical analyses of these possible differences between cultural settings – analyses that are still largely lacking. Moreover, the way in which inter-firm relations are measured and the level of analysis (the single inter-firm relationship, the network of the focal firm, or the entire network within a region) are crucial to developing these understandings. These empirical analyses are, however, fraught with problems, not least because firms that innovate may, for example, have actively searched for embedded relationships in the past. Under these circumstances, is embeddedness a cause or an effect?

In sum, we think that the concept of embeddedness needs to be explored more deeply, both theoretically and empirically. This can only be achieved through the intense exchange of information between researchers in the field. In other words, there is a need for strong ties as we pursue a fuller understanding of the economic significance of strong ties.

34 *Embedded Enterprise and Social Capital*

References

Asheim, B. (1996), 'Industrial districts as "learning regions": a condition for prosperity', *European Planning Studies*, vol. 4, no. 4, pp. 379-400.

Audretsch, D.B. and Stephan, P.E. (1996), 'Company-scientist locational links: the case of biotechnology', *American Economic Review*, vol. 86, pp. 641-652.

Bolton, R. (1998), *A Critical Examination of the Concept of Social Capital*, Paper presented at the Meeting of the Association of American Geographers, Boston.

Boschma, R.A. (1999a), *Culture of Trust and Regional Development: an Empirical Analysis of the Third Italy*, Paper presented at the Congress of the European Regional Science Association, Dublin.

Boschma, R.A. (1999b), 'Learning and regional development', *GeoJournal*, vol. 49, no. 4, pp. 339-343.

Boschma, R.A. and Lambooy, J.G. (1999), 'Evolutionary economics and economic geography', *Journal of Evolutionary Economics*, vol. 9, pp. 411-429.

Camagni, R. (ed) (1991), *Innovation Networks, Spatial Perspectives*, Belhaven Press, London and New York.

Coleman, J. (1990), *Foundations of Social Theory*, Harvard University Press, Cambridge MA.

Dei Ottati, G. (1994), 'Trust, interlinking transactions and credit in the industrial district', *Cambridge Journal of Economics*, vol. 18, pp. 529-546.

Fukuyama, F. (1995), *Trust, The Social Virtues and the Creation of Prosperity*, Hamish Hamilton, London.

Gambetta, D. (1988), *Trust: Making and Breaking of Cooperative Relations*, Blackwell, Oxford.

Grabher, G. (1993), 'Rediscovering the social in the economics of interfirm relations', in G. Grabher, (ed.), *The Embedded Firm. On the Socioeconomics of Industrial Networks*, Routledge, London and New York, pp. 1-31.

Granovetter, M. (1985), 'Economic action and social structure: the problem of embeddedness', *American Journal of Sociology*, vol. 91, no. 3, pp. 481-510.

Håkansson, H. and Johanson, J. (1993), 'The network as a governance structure. Interfirm cooperation beyond markets and hierarchies', in G. Grabher, (ed), *The Embedded Firm. On the Socioeconomics of Industrial Networks*, Routledge, London and New York, pp. 35-51.

Harrison, B. (1992), 'Industrial districts: old wines in new bottles', *Regional Studies*, vol. 26, pp. 469-483.

Hausmann, U. (1996), *Neither Industrial District nor Innovative Milieu: Entrepreneurs and their Contexts. An Actor-oriented Framework and Case Studies from Greater London and Zurich*, Paper presented at the 36th European Congress of the Regional Science Association, Zurich.

Hudson, R. (1999), 'The learning economy, the learning firm and the learning region. A sympathetic critique of the limits of learning', *European Urban and Regional Studies*, vol. 6, no. 1, pp. 59-72.

Lambooy, J.G. (1997), 'Knowledge production, organisation and agglomeration economies', *GeoJournal*, vol. 41, no. 4, pp. 293-300.

Lundvall, B.A. (1993), 'Explaining interfirm cooperation and innovation. Limits of the transaction-cost approach' in G. Grabher, (ed), *The Embedded Firm. On the*

Socioeconomics of Industrial Networks, Routledge, London and New York, pp. 52-64.

Maskell, P. and Malmberg, A. (1999), 'The competitiveness of firms and regions. Ubiquitification and the importance of localized learning', *European Urban and Regional Studies*, vol. 6, no. 1, pp. 9-25.

Morgan, K. (1997), 'The learning region: institutions, innovation and regional renewal', *Regional Studies*, vol. 31, no. 5, pp. 491-503.

Nooteboom, B. (1999), *Inter-firm Alliances. Analysis and Design*, Routledge, London.

North, D.C. (1990), *Institutions, Institutional Change and Economic Performance*, Cambridge University Press, Cambridge.

Oerlemans, L.A.G. (1996), *De Ingebedde Onderneming: Innoveren in Industriële Netwerken*, Tilburg University Press, Tilburg.

Oinas, P. (1997), 'On the socio-spatial embeddedness of business firms', *Erdkunde*, vol. 51, pp. 23-32.

Oinas, P. (1998), *The Embedded Firm? Prelude for a Revived Geography of Enterprise*, Helsinki School of Economics and Business Administration, Helsinki.

Polanyi, K. (1944), *The Great Transformation*, Beacon Press, Boston.

Putnam, R.D. (1993), *Making Democracy Work: Civic Traditions in Modern Italy*, Princeton University Press, Princeton.

Saxenian, A. (1994), *Regional Advantage. Culture and Competition in Silicon Valley and Route 128*, Harvard University Press, Cambridge MA.

Schutjens, V. and Stam, E. (2000), *The Evolution and Nature of Young Firm Networks: a Longitudinal Perspective*, Paper presented at the Symposium on Entrepreneurship, Firm Growth and Regional Development in the New Economic Geography, Uddevala.

Shapiro, C. and Varian, H.R. (1999), *Information Rules, A Strategic Guide to the Network Economy*, Harvard University Press, Harvard.

Storper, M. (1997), *The Regional World: Territorial Development in a Global Economy*, Guilford Press, New York.

Taylor, M. (1999), *Entreprise, Embeddedness and Exclusion: Buyer-supplier Relations in a Small Developing Country Economy*, Working Paper, University of Portsmouth, Portsmouth.

Uzzi, B. (1996), 'The sources and consequences of embeddedness for the economic performance of organizations: the network effect', *American Sociological Review*, vol. 61, pp. 674-698.

Uzzi, B. (1997), 'Social structure and competition in interfirm networks: the paradox of embeddedness', *Administrative Science Quarterly*, vol. 42, no. 1, pp. 35-67.

Uzzi, B. (1999), 'Embeddedness in the making of financial capital: how social relations and networks benefit firms seeking financing', *American Sociological Review*, vol. 64, pp. 481-505.

Zukin, S. and Di Maggio, P. (eds) (1990), *The Social Organization of the Economy*, Cambridge University Press, Cambridge.

Chapter 3

Rethinking Institutions and Embeddedness in a Third World Context

Giles Mohan

Introduction: Disappointing Geographies, Limited Economies

> The challenge for development geography as a profession is in our view to apply the new knowledge generated about the First World growth processes for use in the Second and Third World. (Reinert and Risser 1995, p. 30)

> By throwing the cover of culture over material relationships, as if one had little to do with the other, such a focus diverts criticism of capitalism to the criticism of Eurocentric ideology … [and] … provides an alibi for inequality, exploitation, and oppression in their modern guises. (Dirlik 1994, p. 346)

At the broadest level, this chapter is concerned with a seemingly entrenched division in geography: the perceived and effective difference between development geography and economic geography. In spite of some useful debates around interconnectedness and transgression (Slater 1992), economic geography remains largely untouched by a counter-critique from the Third World. The over-riding impression left by collections such as Lee and Wills' (1997) *Geographies of Economies* is that what constitutes legitimate economic geography, even with its widened, 'new' agenda, does not include the Third World.

This chapter begins to address this lacuna, although this is not simply a cathartic exercise in order to 'right wrongs'. In this chapter, as in the other contributions to this collection, I contend that much can be learned about economic growth by studying economic processes outside the countries of the core. As globalisation intensifies, local, regional and national economic spaces become increasingly inter-linked which opens up the possibility, and likelihood, that similar processes operate across space.

This is not to suggest an inevitable convergence, but to open up a possible dialogue between experiences in different places. I begin this chapter by examining two critical perspectives in human geography – eurocentrism and economism – that help define and delimit the sub-disciplines of economic geography and development studies. In the next section, I demonstrate the implications of this division through an analysis of institutional and social capital theories that, in turn, permeate policy discourses. It shows how the uncritical transfer of ideas from 'core' to 'periphery' reflects deeper ideological movements under the current regime of reformist neo-liberalism. After arguing that such bifurcations are intellectually unhelpful and ideologically loaded, I move on to examine key processes that are at work in both 'developed' and 'developing' areas. I focus on questions of embeddedness and clustering which are generally thought to enhance firm and place competitiveness. By studying these processes outside of core economies we get a more complex picture regarding the social and political relations that underpin them. In particular, that embeddedness and inter-firm collaboration can be a means of coping with the intense economic strains of poverty rather than a more proactive strategy of regional competitiveness. I conclude by suggesting where these tentative insights might take us in the future.

Eurocentric Economism

My current concerns are rooted in the two intertwining heritages of eurocentrism and economism. Recently, the eurocentric nature of human geography has been challenged (Slater 1992). Shohat and Stam (1994) argue that eurocentrism as a discourse is, 'naturalized as "common sense"' (p. 1) and, 'sanitizes Western history while patronizing and even demonizing the non-West' (p. 3). Slater (1989, 1992) echoes this in his reading of various critical texts in human geography. He states that, 'there is a pervasive tendency to construct research agendas, formulate key issues for theoretical debate, and draw on bodies of literature as if the West were somehow a self-contained entity' (Slater 1992, p. 309). Even where theorists in economic geography have challenged occidental exceptionalism, they do so through examples drawn from successful Pacific Rim economies (Thrift and Olds 1996).

In contrast, researchers in development studies have for some time been aware of the economism of their concepts (Leys 1996, Corbridge 1986, Brohman 1995). For example, Leys (1996 [originally 1977]) criticised the ways in which dependency theorists borrowed uncritically from neoclassical economists and in so doing equated peripheral development with metropolitan capitalism. It is only recently that economic geographers have become reflexively aware of the economism of their discipline (Gibson-Graham 1995, Thrift and Olds 1996). This is in part through the challenges of gendered accounts of work and the more recent cultural and sociological turns which stress the cultural embeddedness of economic practices (Granovetter and Swedberg 1992, Crang 1997). However, there is still a pervasive privileging of an authentic 'economy' within mainstream economic geography that has far broader implications. As Gibson-Graham notes, '[c]apitalism's others fail to measure up to it as the true form of economy ... non-capitalist forms of economy often present themselves as a homogenous insufficiency rather than as positive and differentiated others' (1995, p.278). In transcending this they argue, as do Thrift and Olds (1996), that we should recognise economic plurality.

However, we need to see how these two lineages are constitutive of each other. In general, defining legitimate economic activity as capitalist relates back to industrialisation in Europe when development became equated with processes of capitalism and modernity within the core economies (Escobar 1995, Cowen and Shenton 1996). It then becomes tautological for eurocentric, western-based academics to see this as a spatially delimited, authentic form of economic development whose absence creates the obvious and inevitable need for those regions outside to 'catch up'. Anything falling outside this self-referential lens is considered 'other' and illegitimate, and unworthy of the label 'economy'. As Gibson-Graham (1995) notes, 'Non-capitalist forms are thus to be found in so-called Third World countries that lack the fullness and completeness of capitalist development' (p. 278, also see Latouche 1994). The implications of this relate to the ways in which GDP/GNP calculations wholly misrepresent actual economic processes based around subsistence and reciprocity or the ways in which economic activities are labelled as being 'informal', 'black', 'non-tradable' or 'shadow'.

Eurocentrism and economism play into one another so that ontologically, spatially and academically development geography and economic geography are treated as separate domains. This sets up a paradox whereby the 'normal' capitalist economy is privileged vis-à-vis its 'other', giving the impression that the two are not functionally linked and/or that the same analysis is not amenable to such 'different' spheres. Where analytical ideas and policies do pass between them it is usually a case that development geographers borrow uncritically from ideas first applied to the authentic, western economic experience or, worse still, these ideas are forced as policies on developing countries through various forms of 'conditionality'. However, I firmly support the need for exchanges between these two areas if both are to be enriched.

Eurocentrism, Economism and Conditionality: New Institutionalism, Social Capital and the Good Governance Agenda

The purpose of this section is to briefly demonstrate the implications of eurocentrism and economism in recent academic-policy discourses. Since neo-imperialism takes many guises, I have chosen to examine the transfer of intellectual ideas from 'north' to 'south'. Such transfers mirror the financial flows of loans and aid and are, in effect, a subtle form of conditionality. I focus on theoretical discussions of institutions that have been gaining in importance in development thinking and policy over the past decade (Toye 1993, Leys 1996, Hyden 1997, Harriss and de Renzio 1997, Fine 1999). These 'new institutionalist' and 'social capital' models form the logical underpinnings of much of the orthodox governance debate (Leftwich 1994) on how to promote economic development in the Third World. The spread of these perspectives in development theory and policy has been a recent phenomenon, but is of growing importance. The World Bank regularly cites the works of Robert Bates and Robert Putnam even though there are relatively few empirical applications of these theories in a development context (Nabli and Nugent 1989, Harris and de Renzio 1997).

The New Institutionalism and Third World Economic Development

The New Institutionalism comprises a number of related branches that share many of the rationalising assumptions of neo-classical economics, but builds upon them by exploring the ways in which market failure occurs. They begin by arguing that the individualism of *homo oeconomicus* in neo-classical economics does not exist. Instead they seek to understand why markets fail by extending the concept of rationality into examining institutions. As Bates (1993, p. 1) states the new institutionalism, 'attempts to apply to non-market institutions the same form of reasoning that neo-classical economists have applied to the analysis of markets'. Within this general focus on the 'non-market' there are two inter-related strands of theorising. First, transaction costs and imperfect information and, second, collective action (Williamson 1985, Bardhan 1989, Nabli and Nugent 1989, Leys 1996).

If we explore the work of Robert Bates, who has worked on African policy matters and is considered an important social theorist, we find a paradoxical view of the rational individual. Bates' ontology begins with the existence of ranked preferences and he argues that, 'rationality merely requires that the individual decision makers choose his or her "top" or most preferred alternative ... [although] ... rationality says nothing about the content of preferences' (1987, pp. 134-135). Bates stresses that by positing a range of preferences he allows for various motivations and, therefore, is not being culturally biased. However, when we explore his work we find that this range of preferences is reduced to stark either/or choices. For example, in his work on the Nuer, he argues that peasants reduce long-term risk by either investing in cattle *or* in education. This dichotomy does not reflect a range of preferences, and also equates rational choice only with income decisions. In Bates' work, therefore, we are presented with an individual who is, to all intents and purposes, outside any specific cultural context.

Bates develops his argument such that the, 'preferences of individuals aggregate into choices for society' (1987, p.136) and it is 'politics' and 'institutions' that act as the channels of aggregation (Peters 1993). However, in much of the literature, institutions are never clearly specified. In some cases they are 'organic' and relate to social rules such as the market, whereas in others they are more 'pragmatic' and emerge to fulfil specific functions (Bardhan 1989). The question of how these institutions form is, however, never addressed. As Bardhan notes, 'An institution's mere function of serving the interests of potential beneficiaries is clearly inadequate in *explaining* it' (1989, p. 1390, original emphasis). For example, if institutions serve the social good why would they emerge if 'rational' individuals were aware of potential free-rider issues? These functionalist accounts of institutions cannot overcome this tension between individual and social benefits.

In contrast to the stated methodological individualism in much rational choice theory (RCT), the focus is more often than not on interest groups, coalitions and various social strata (Stein and Wilson 1993). Where Bates discusses individuals they tend to be reduced to the characteristics of the group from which they are drawn (Peters 1993). Such analytical problems extend into analysing the major institution that distorts the market – the state. Kreuger (1974) argues that state intervention creates rents that are competed over in much the same way as contracts in a free market. Bates' work builds on this foundation by arguing that all state actors are motivated by personal gain either in the form of rent-based income or simply to hold on to power. Despite his claim to be more enlightened than neo-classical economists, Bates reduces the activities of state actors to simplistic calculi. For example, agricultural subsidies simply buy off larger farmers in order to prevent them forming oppositional coalitions. In the same way as peasants are 'a-cultural' these state actors have no other possible motivations such as ethnicity, religion, or nationalism.

Bates' a-structural, a-social analysis is double-edged and also requires an analysis of political economy. For what Bates does not do, and his followers in the donor institutions cannot do, is acknowledge that this cultural and economic essentialism is bound up in their geo-political imagination. Attempts to 'naturalise' this form of subjectivity are 'by no means a "natural" starting point, as Bates implies - it is a highly political one that takes as "natural" the very issue that is at stake in the struggle for Africa's future' (Leys 1996, p. 52). Such analyses run the risk of ignoring the complex and reciprocal linkages between international political economy and national socio-political formations and, in doing so, obfuscate the massive use of state power by the institutions of global governance to create 'free markets'.

Social Capital and Regional Growth

Social capital models differ from rational choice institutionalism by focusing more strongly on political culture (Hyden 1997), though both discourses share similar views on the interaction of local society and institutions. Concepts of social capital have long roots in social science, but have been revived most famously by Robert Putnam (1993, 1995) in his work on Italy and America. For Putnam, social capital comprises, 'features of social organization, such as networks, norms, and trust, that facilitate coordination and cooperation for mutual benefit' (Putnam 1993, p. 35). This, in turn, fosters reciprocity and facilitates information flows and is, in essence, the socio-cultural 'glue' that binds communities together and ensures both political and economic progress. At the level of macro-policy, the World Bank sees social capital underpinning a wide range of developmental initiatives. On their social capital website (www.worldbank.org/poverty/scapital/index.htm), social capital is interpreted as contributing to economic prosperity and sustainability and should ideally entail horizontal and vertical associations which promote social cohesion on the one hand and prevent divisive parochialisms on the other. In terms of Third World industrial organisation, academics have recently suggested a positive role for social capital in regional collaboration (Schmitz 1998, Perez Sainz 1998, McCormick 1999).

An extensive critique of social capital is beyond the scope of this chapter (see Harriss and de Renzio 1997, Hyden 1997, Fine 1999), but it is worth mentioning some of the problems concerning its reading of culture and political economy. First, in Putnam's work we see an internalist view of culture and politics. Successful regions are those that have had high stocks of social capital for a long time, locking them into a 'path dependency' in which those initial stocks of social capital initiate a self-reinforcing cycle of prosperity. The corollary is that unsuccessful regions lacked and lack the 'proper' type or stock of social capital. So, while the concept of social capital promises to provide a more nuanced understanding of local development, it actually reduces causality to a rigid determinism. This allows the proponents of social capital, especially the major lenders, to sidestep the problems of the state and its relation to the global economy, since economic considerations are rendered unproblematic. It is simply the shortcomings of local society that prevent it from inserting itself into a virtuous cycle of economic growth. As with modernisation theory, the failure to develop rests with the 'victims' of poverty and under-development and not the wider political economy.

Leading from this cultural internalism is the problem that Putnam in particular ignores – the role of the state in enabling or destroying social capital. Tarrow's (1996) divergent reading of Italian political history shows that the Southern regions were held in a semi-colonial relation to the North so that economic development and associational life were suppressed. As such, the contemporary economic weaknesses and the social capital deficit of those Southern regions are not a reflection of initial endowments. Similarly, Putzel (1997) adds that in areas identified as having high stocks of social capital, other political organisations were heavily involved. In particular in the North, the communist trade unions were active in promoting political activism and civic life in general. It is this lack of attention to the state's role and the roles of other forms of political organisation that prompted Evans (1997) and Ostrom (1996) to talk of state-society synergy.

This privileging of certain forms of social capital over others reflects the broader trend of economism and eurocentrism. Whether tacit or overt, social capital theory attempts to strengthen economic growth and in doing so attempts to give an economic rationale to all 'non-economic' behaviour. Although never clearly stated, the very notion of 'capital' in 'social capital' consigns it to the more general category of capital. Unlike stock resources, which diminish with exploitation, the 'use' of social capital increases its worth and contributes to success. So, again, despite a rhetoric of cultural sensitivity, the '... only proviso is that social capital should be attached to the economy in a functionally positive way for economic performance, especially growth' (Fine 1999, p. 5). As Fine goes on to argue, '... social capital allows the World Bank to broaden its agenda whilst retaining continuity with most of its practices and prejudices which include benign neglect of macro-relations of power' (1999, p. 12).

Both economistic and eurocentric approaches to culture, institutions and development under-theorise the state and wider political-economy. They reify an internalised view of culture, and are unable to demonstrate causal processes in space and time. Given these severe limitations we can only conclude that these theories, in Robert Wade's terms, secure 'paradigm maintenance' which attempts to bring out the *homo oeconomicus* in all of us (Williams 1999). In this sense, we must be wary of uncritically transferring models developed in particular political and economic contexts without being attentive to the ideological processes underpinning their transfer.

Beyond Eurocentrism and Economism: the Possibilities of Embeddedness

This section moves the debate beyond the economistic and eurocentric readings of the economy. The assumption of many researchers is that processes operating in developing and developed regions are place-specific. However, I want to show where similarities lie between regions and how exchanges between seemingly discrete experiences can engender new insights in both. This research lies in the area of industrial districts and clustering (Sengenberger and Pyke 1992, Rasmussen *et al.* 1992, Schmitz 1993) which feed into broader debates around embeddedness, flexible specialisation and regional development (Granovetter and Swedberg 1992).

Much of this work takes on board the insights of anthropological and sociological studies of 'economic' behaviour (Granovetter 1985, Zukin and DiMaggio 1993, Smelser and Swedberg 1994). These debates first emerged in the mid 1980s through such things as the localities initiative, but have only recently risen to prominence among policy-oriented economic geographers (Amin and Thrift 1994, Cooke and Morgan 1991, Clark *et al.* 2000). At one level, they seek to theorise and study the importance of the 'non-economic' dimension of economic growth that is promising in challenging the economism that infects much theorising of economic development (Crang 1997). For me this is its greatest use and, in a developing country context, provides useful ammunition to counter the excessive economism of neo-liberal approaches.

It is problematic to synthesise the vast literature on clustering and industrial districts, but the argument revolves around the networked relationships between small and medium sized firms that operate collaboratively to achieve regionally-based flexibility which, in turn, successfully serves the demands of fragmented, but globalising, markets (Schmitz and Musyck 1993, Cooke 1996). In accounting for this success, researchers have turned to various sociological explanations centred on local culture and institutional collaboration. Amin and Thrift (1994) argue that contrary to the implication that globalisation is a universal and homogenising process it necessarily takes 'place' through the 'local', and they see a return to the region as the, '... basic unit of economic, cultural, and political organisation, as a result of the crisis of the national state-based system of capital accumulation' (p. 7). We then find a discussion of 'successful' European regions that find themselves at the '... centre of global economic circuits' (p. 10). Such a discussion picks up on much of the work on industrial districts and economic sociology that stresses the importance of 'institutional thickness' in determining economic success.

Cooke (1997) echoes this background by arguing that these 'intelligent regions' possess 'institutional reflexivity' which represents a new form of post-modern economic governance. For Cooke, these regions are marked by their collaborative qualities that promote social justice on their so-called 'high road' to economic growth (Sengenberger and Pyke 1992). In these models there is a general movement towards increasing institutionalisation and collaboration (Schmitz and Musyck 1993, Cooke and Morgan 1991) that facilitates spontaneous regional growth. And, as Cooke notes, President Clinton at the Detroit meeting of the G7 in 1994 suggested that this model of economic development should be the blueprint for solving the global economic crisis.

However, a general criticism of these models relates to Moon and Prasad's (1994) accusation in the East Asian context that, '[b]eing preoccupied with the explanation of "successful" outcomes, both statist and network perspectives have produced an incomplete portrayal of the ... political economic landscape. ... No longer should analysis be confined to success. The ontological horizon should be expanded to include both successes and failures' (p. 376). By anchoring policy prescriptions in a few, possibly unique and exceptional cases, we run the risk of cementing unrealistic and unrealisable policy prescriptions. It is important to examine other contexts where clustering and regionally based collaboration occurs to determine whether, indeed, embeddedness and reciprocity are so universally positive.

Embeddedness, Clustering and Informality

The flexible specialisation and industrial districts literatures argue that collective efficiency can be a competitive asset which overcomes the limitations imposed by the lack of scale economies. It also promotes learning within and between firms leading to enhanced innovation and thereby competitive success. Further, the flexibility that such collaboration makes possible can also promote competition as it frees smaller firms from the rigidities of integrated firm structures. In a developing country context, debates about the role of small firms in regional economic growth overlap with a large body of literature on the 'informal sector' (Portes *et al.* 1989, Ypeij 1998, Perez Sainz 1998). It is in the interstices and overlaps between these two bodies of literature that we get a sense of what 'unsuccessful' regional networks might look like. I do not propose to review the theoretical debates surrounding 'informality' (see Castells and Portes 1989, Thomas 1995, Rakowski 1994, Perez Sainz 1998), but rather to use these alternative studies of industrial organisation to question the universalism of the 'clustering' and 'industrial districts' models built on principles of flexible specialisation.

One important trend in the informal sector is that rather than the sector being remnants of 'traditional' or 'pre-capitalist' economies, which diminish as industrialisation proceeds, there has been a growth in its size and importance. Following the restructuring and debt crisis of the 1970s, liberalisation programmes have ushered in civil service retrenchment, privatisation and the slimming down of state-owned enterprises, enhanced competition from large foreign firms, the removal of subsidies (which have a tendency to depress real wages), and various exchange rate adjustments, which have all served to push more people into relying on a harsh mixture of 'petty' economic activities (Thomas 1995, Ypeij 1998). As Castells and Portes (1989) observe the 'space' opened up to the informal sector in a national economy varies, 'according to the pace of social unrest and the political orientations of government' (p. 26). Crucially, the informal sector is not confined to LDCs, but is alive and growing in most developed countries, as labour market conditions have tightened (Castells and Portes 1989, Rakowski 1994).

In an examination of the dynamics of small firm clusters in LDCs and the role of the informal sector, it is important to differentiate between different processes and types of cluster. Perez Sainz (1998) sees such differentiation assuming three forms: (1) exclusion from globalisation, which forces the extreme poor into various survival or coping strategies; (2) limited inclusion in globalisation but in a subordinate position; and (3) inclusion in globalisation on a more dynamic and proactive basis.

These different sets of processes allow us to conceive of informality and small firm networks in quite different ways. In the first type of cluster, the process of 'atomisation' dominates, and there is little interaction between economic operators. The term 'firm' does not apply to these petty traders and other marginal and informal workers. Here household resources, including labour, are pooled. Hence, cooperation between agents is primarily a survival strategy or coping mechanism. In the second type of cluster, informality functions as a form of disguised wage labour (Rakowski 1994) – processes that had previously been identified by a Marxist school. Larger firms serving national or international markets reduce both costs and the threats of labour organisation by decentralising production to various subcontractors who, in turn, put out production to tiny, often informal, producers such as home-workers. In the third type of cluster, the full spectrum of place-specific inter-firm relationships and processes dependent on proximity nurture dynamic regional growth. This is the 'industrial district' of the developed world. While they are tied into global manufacturing and marketing webs, these districts have achieved power through inter-firm collaboration.

A number of studies demonstrate and confirm the different scenarios suggested above (Perez Sainz 1998, Ypeij 1998, McCormick 1999, Lanzetta de Pardo *et al.* 1989, Dawson 1992). However, the focus of this chapter is on the industrial clustering model. The first scenario involving informal activity as a survival strategy will not be examined since it is essentially individualistic and presents only limited possibilities for collaboration. The second scenario on dependent inclusion in global markets is similar to McCormick's (1999) two types of clusters which either lay the groundwork for industrialisation or show signs of more sustained, cooperative industrialisation. In this scenario, subcontracting is an important element (Aeroe 1992). Ypeij's (1998) study of Lima shows that two levels of subcontracting exist. The first is *non-equivalent* subcontracting that takes place between firms of different sizes where a process stage in large batch production (in this case stitching shoes) is contracted to a number of small firms. The second form of subcontracting is *equivalent* in the sense of taking place between firms of similar size. While the former inter-firm relationships are based principally on costs rather than quality (and other lesser criteria), *equivalent* subcontracting, '... takes place between producers who know each other well through their social networks of friends, family members, and neighbours' (Ypeij 1998, p. 87). Dawson (1992) has observed similar inter-firm relationships in a study of small engineering firms in Ghana, and by Sverrisson (1992) in a study of woodworking in Kenya and Zimbabwe, though often in response to limited technological capacity. Collaboration either allowed a number of firms with a limited technical division of labour to produce 'bespoke' products by carrying out different stages of a production process or, more commonly, involved firms simply loaning each other tools. In such clusters, the level of joint action via formalised institutions was limited (McCormick 1999). The third scenario involves the more complex and proactive processes of 'clustering' in industrial districts. Since this is the main focus of this chapter, this industrial clustering model, in which collaboration between firms and entrepreneurs tends to be more formalised (Rasmussen *et al.* 1992, McCormick 1999), will be explored in more detail in the next section of the chapter.

Industrial Districts: Limited Cooperation

One of the general trends identified in the industrial districts literature is that institutionalisation of cooperation increases over time in such areas as training and the provision of 'real services' (Schmitz and Musyck 1993, Cooke, 1997). Allied to this is an assumption that transactions between firms and between firms and institutions are based on trust and reciprocity with a level of regional collaboration and coherence that amounts to the 'socialisation of risk'. In this regard, 'socio-territoriality' and proximity are important as regular, face-to-face interaction strengthens inter-firm relations. Certainly, this culturally embedded and institutionally strengthened analysis of regional economies has certain validity, but is it entirely accurate?

Empirical material suggests that this neat teleology breaks down. In a study of the Brazilian footwear industry in the Sinos Valley 'industrial district', Schmitz (1993) argues that cooperation fell apart in the face of global competition. Shoe producers lowered factory gate prices as information on Korean shoe prices reached them via marketing agents. Hence, price-based, inter-firm competition within the industrial district became intense and this was passed on to workers through wage squeezes and the lowering of working conditions. During this time, '… socio-cultural ties weakened and had less of an influence on inter-firm relationships' (p. 27). A subsequent study (Schmitz 1998) showed that after a period of non-cooperation in the late 1980s and early 1990s moves were made to upgrade the quality of the Brazilian shoe industry as Chinese and Vietnamese producers had captured an increasing share of the cheap, mass shoe market. In this sense, the need to cooperate increased, '… not because of socio-cultural ties but because of the economic costs of not cooperating' (Schmitz 1993, p. 27). However, recent collective efforts around export schemes such as the 'Shoes from Brazil Programme', organised by a large manufacturers association, were unsuccessful because the five largest firms in the Sinos Valley failed to comply. Schmitz (1998) argues that, '... some of the largest enterprises ... had integrated vertically and thus reduced their dependence on the cluster' (p. 31) and, '... collaborated very closely amongst themselves, practising openness towards each other but being inaccessible to outsiders' (p. 33). This shows that larger, more powerful firms felt that complex subcontracting webs, contrary to the industrial district theories, compromised quality and flexibility so that it was more efficient to integrate production functions by increasing plant size and reorganising around teams rather than traditional production lines. Having integrated and improved quality, these firms sought to protect their powerful position by working towards limited cooperation with like-sized firms rather than participating in region-wide collaboration.

Similar processes of weakening of socio-cultural ties are evident in other studies. In a study of Costa Rican artisans, Perez Sainz (1998) observes that, '... competition seems to prevail over cooperation' (p. 174) because competition is not based on product specialisation or quality, but largely on price. Newcomers rarely innovate, but simply copy the products of existing artisans and attempt to undercut them. In the context of ethnic businesses in New York, Waldinger (1995) observes that in terms of human capital, cultural ties of ethnicity did not always count as much as 'objective' qualifications, although Castells and Portes (1989) suggest that new migrants tend to rely on fellow ethnic group members through a feeling of threat. However, such ties weaken as individuals, families and firms become more established. These examples all suggest that economic competition and market rationality still prevail despite the culturally embedded ties of trust and reciprocity. This is not to deny that such local ties are important, but to argue instead that their existence is fluid and reversible and that a stronger agency-centred analysis is needed. These cases also suggest that we should be attentive to the operation of power relations between firms.

Hence, another reading of the progressive institutionalisation within industrial districts that has not been adequately explored is that 'institutional thickness' may be a way of 'locking out' outsiders rather than proactively facilitating insiders within a region. As White (1993) observes, '[s]uccessful market closure undertaken through the collective action of market participants results in the establishment of conditions which protect or extend the market position of those actors' (p. 7). This brings us to the question of the actual functional operation of these institutions rather than observations of their form. There seems to be a certain amount of inferred causality or 'reading off' from the embeddedness model. In discussing the actual role played by various institutions, Schmitz and Musyck (1993) argue that, '[t]he mere existence of such services does not mean that they function well or are effective, although this is often implied in the literature' (p. 20). No real evidence is provided for how these institutions actually socialise risk. In the end, we are left with a form of functionalism whereby institutions provide the 'glue' that enables economic growth. At no point do we go beyond institutions as homogenous totalities that cooperatively act in the best interests of the regional whole. These are assumptions that must be questioned.

Conclusion

The previous section demonstrated that the recent focus on embeddedness in economic geography and development studies has the potential to complicate and enrich our understanding of regional growth processes in an increasingly globalised world. In particular, it challenges economism, is more attuned to cultural differences, and can produce detailed analysis of processes in particular places. However, there are certain caveats that must be attached to this conclusion of which three are particularly important.

First, in promoting an 'internalist' explanation of relative economic performance, we do not see much consideration of other levels of economic governance. In reifying the 'local' and its socio-cultural specificity, we are presented with an internalised story of economic success. However, as Schmitz and Musyck (1993) state, '... the effect of the macro policy environment on the development of small- and medium- sized enterprises is not dealt with in the industrial district literature. This automatically disregards the explanatory power of the national macro economic environment for the relative economic performance of industrial districts' (p. 33). As with the social capital literature, the role of the state, in particular, is relatively under-privileged as Markussen (1999) forcibly demonstrates with respect to the role of military spending in the success of Silicon Valley.

Second, something that emerges from the rational choice institutionalism, social capital and industrial district literatures is their mis-specification of the roles played by institutions. I would argue that the empirical analysis based upon these schools of thought largely infers causality from a pre-given model. Certain institutions and relations clearly exist, but there is little convincing evidence that they formed to fulfill the role expected of them in theory. As Peters (1993, p. 1071) asserts, '[c]asting institutions in this way pre-empts analysis of their specific structure and organisation, of the principles, ideas and values informing their organization, of the connections across institutions, and of the way different social categories of persons, as well as individual actors, affect and are affected by different institutions'. What is needed is a closer analysis of the formation and operation of actually existing institutions and the potential these have for locking out competitors within a region.

Third, if we follow Gibson-Graham's (1995) call for pluralising 'the economic' so that, 'noncapitalism is released from its singularity and subjection, and becomes potentially visible as a differentiated multiplicity of economic forms' (p. 279; see also Latouche 1994), we must avoid reifying and celebrating poverty. In a reversal of the economism of the neo-liberals, this approach can dissolve economic difference into a dangerous relativism where any form of activity has counter-hegemonic potential. However, as Campbell (1997) notes with respect to some western approaches to sustainable development, '… interpreting African dire necessity as a product of "indigenous knowledge" rather than a product of grinding poverty, the concept of indigenism can then be served up to gullible Westerners as a "sustainable" system that they should be proud to live by' (pp. 50-51; see also Brass 1995). If economic plurality is to be analytically and politically meaningful, then there must be some normative criteria by which certain forms of 'other' economic activity are deemed unsustainable. As the analysis of restructuring and coping showed such 'informal' activities are linked to the dynamics of global capitalism and grand schemes of liberalisation. Informality may, for example, be an inevitable and unavoidable consequence of the erosion of trade union power so that workers are forced into even more precarious situations. Unless these processes of 'othering' are more critically examined we run the risk of popularising an anti-economism almost as damaging as the economism it seeks to subvert.

References

Aeroe, A. (1992), 'New pathways in industrialization in Tanzania: Theoretical and strategic perspectives', *IDS Bulletin*, vol. 23, no. 3, pp. 15-20.
Agnew, J. and Corbridge, S. (1995), *Mastering Space: Hegemony, Territory and International Political Economy*, Routledge, London.
Amin, A. and Thrift, N. (eds) (1994), *Globalisation, Institutions and Regional Development in Europe*, Oxford University Press, Oxford.

Bardhan, P. (1989), 'The new institutional economics and development theory: A brief critical assessment', *World Development*, vol. 17, no. 9, pp. 1389-1395.

Bates, R. (1987), *Beyond the Miracle of the Market*, Cambridge University Press, Cambridge.

Bates, R. (1993), *Social Dilemmas and Rational Individuals: An Essay on the New Institutionalism*, Paper presented to conference of the Third World Economic History Group, 17-19 September, Department of Economic History and the Development Studies Institute, University of London.

Booth, D. (1985), 'Marxism and development sociology: Interpreting the impasse', *World Development*, vol. 13, no. 7, pp. 761-787.

Brass, T. (1995), 'Old conservatism in "new" clothes', *The Journal of Peasant Studies*, vol. 22, no. 3, pp. 516-540.

Brohman, J. (1995), 'Economism and critical silences in development studies: a theoretical critique of neoliberalism', *Third World Quarterly*, vol. 16, no. 2, pp. 297-318.

Campbell, A. (1997), *Western Primitivism: African Ethnicity*, Cassell, London.

Castells, M. and Portes, A. (1989), 'World underneath: the origins, dynamics, and effects of the informal economy', in A. Portes, M. Castells and L. Benton (eds), *The Informal Economy: Studies in Advanced and Less Developed Countries*, Johns Hopkins University Press, Baltimore, pp. 11-35.

Clark, G., Feldman, M. and Gertler, M. (eds) (2000), *The Oxford Handbook of Economic Geography*, Oxford University Press, Oxford.

Cooke, P. (1996), 'Reinventing the region: Firms, clusters and networks in economic development', in P. Daniels and W. Lever (eds), *The Global Economy in Transition*, Longman, Harlow, pp. 310-327.

Cooke, P. (1997), 'Institutional reflexivity and the rise of the region state', in G. Benko and U. Strohmayer (eds), *Space and Social Theory*, Blackwell, Oxford, pp. 285-301.

Cooke, P. and Morgan, K. (1991), 'The intelligent region', *Regional Industrial Research Report*, no. 7, CASS, Cardiff.

Corbridge, S. (1986), *Capitalist World Development*, Macmillan, London.

Cowen, M. and Shenton, R. (1996), *Doctrines of Development*, Routledge, London.

Crang, P. (1997), 'Introduction: Cultural turns and the (re)constitution of economic geography', in R. Lee and J. Wills (eds), *Geographies of Economies*, Arnold, London, pp. 3-15.

Dawson, J. (1992), 'The Relevance of the Flexible Specialisation Paradigm for small-scale industrial restructuring in Ghana', *IDS Bulletin*, vol. 23, no. 3, pp. 34-38.

Dirlik, A. (1994), 'The postcolonial aura: Third World criticism in the age of global capitalism', *Critical Inquiry*, vol. 20, Winter, pp. 328-356.

Escobar, A. (1995), *Encountering Development: The Making and Unmaking of the Third World*, Princeton University Press, Princeton.

Evans, P. (1997), 'Introduction: Development strategies across the public-private divide', in P. Evans (ed), *State-Society Synergy: Government and Social Capital in Development*, IAS Research Series No. 94, University of California Press, Berkeley, pp. 1-10.

Fine, B. (1999), 'The developmental state is dead – long live social capital?', *Development and Change*, vol. 30, pp. 1-19.

Fox, J. (1997), 'How does civil society thicken? The political construction of social capital in rural Mexico', in P. Evans (ed.), *State-Society Synergy: Government and Social Capital in Development*, IAS Research Series No. 94, University of California Press, Berkeley, pp. 19-149.

Gibson-Graham, J.K. (1995), *The End of Capitalism (As We Knew It)*, Blackwell, Oxford.

Granovetter, M. (1985), 'Economic action and social structure: The problem of embeddedness', *American Journal of Sociology*, vol. 3, pp. 481-510.

Granovetter, M. and Swedberg, R. (eds) (1992), *The Sociology of Economic Life*, Westview Press, Boulder.

Harriss, J. and P. de Renzio (1997), '"Missing link" or analytically missing? The concept of social capital: An introductory bibliographic essay', *Journal of International Development*, vol. 9, no. 7, pp. 919-937.

Hyden, G. (1997), 'Civil society, social capital, and development: Dissection of a complex discourse', *Studies in Comparative International Development*, vol. 32, no.1, pp. 3-30.

Kaplinsky, R. and Manning, C. (1998), 'Concentration, competition policy and the role of small and medium-sized enterprises in South Africa's industrial development', *The Journal of Development Studies*, vol. 35, no. 1, pp. 139-161.

Kreuger, A. (1974), 'The political economy of the rent-seeking society', *The American Economic Review*, vol. 64, no. 3, pp. 291-303.

Lanzetta de Pardo, M. and Castano, G. with Soto, A. (1989), 'The articulation of formal and informal sectors in the economy of Bogota, Colombia', in A. Portes, M. Castells and L. Benton (eds), *The Informal Economy: Studies in Advanced and Less Developed Countries*, Johns Hopkins University Press, Baltimore, pp. 95-110.

Latouche, S. (1994), *In the Wake of Affluent Society: An Exploration of Post-Development*, Zed Books, London.

Lee, R. and Wills, J. (eds) (1997), *Geographies of Economies*, Arnold, London.

Leftwich, A. (1993), 'Governance, democracy and development in the Third World', *Third World Quarterly*, vol. 14, no. 3, pp. 605-624.

Leftwich, A. (1994), 'Governance, the state and the politics of development', *Development and Change*, vol. 25, pp. 363-386.

Levi, M. (1996), 'Social and unsocial capital: A review essay of Robert Putnam's *Making Democracy Work*', *Politics and Society*, vol. 24, no. 1, pp. 45-55.

Leys, C. (1996), *The Rise and Fall of Development Theory*, James Currey, London.

Markusen, A. (1999), 'Fuzzy concepts, scanty evidence, policy distance: The case for rigour and policy relevance in critical regional studies', *Regional Studies*, vol. 33, no. 9, pp. 869-884.

McCormick, D. (1999), 'African enterprise clusters and industrialization: Theory and reality', *World Development*, vol. 27, no. 9, pp. 1531-1551.

Moon, C. and Prasad, R. (1994), 'Beyond the developmental state: Networks, politics and institutions', *Governance: An International Journal of Policy and Administration*, vol. 7, no. 4, pp. 360-386.

Nabli, M. and Nugent, J. (1989), 'The new institutional economics and its applicability to development', *World Development*, vol. 17, no. 9, pp. 1333-1347.

North, D. (1989), 'Institutions and economic growth: An historical introduction', *World Development*, vol. 17, no. 9, pp. 1319-1332.

Ostrom, E. (1996), 'Crossing the great divide: Coproduction, synergy and development', *World Development*, vol. 24, no. 6, pp. 1073-1087.

Perez Sainz, J. (1998), 'The new faces of informality in Central America', *Journal of Latin American Studies*, vol. 30, pp. 157-179.

Peters, P. (1993), 'Is "rational choice" the best choice for Robert Bates? An anthropologist's reading of Bates's work', *World Development*, vol. 21, no. 6, pp. 1063-1076.

Portes, A., Castells, M. and Benton, L. (eds) (1989), *The Informal Economy: Studies in Advanced and Less Developed Countries*, Johns Hopkins University Press, Baltimore.

Putnam, R. (1993), 'The prosperous community: Social capital and public life', *The American Prospect*, vol. 13, pp. 35-42.

Putnam, R. (1995), 'Bowling alone: America's declining social capital', *Journal of Democracy*, vol. 6, no. 1, pp. 65-78.

Putzel, J. (1997), 'Accounting for the "dark side" of social capital: Reading Robert Putnam on democracy', *Journal of International Development*, vol. 9, no. 7, pp. 939-949.

Rakowski, C. (1994), 'Convergence and divergence in the informal sector debate: A focus on Latin America, 1984-92', *World Development*, vol. 22, no. 4, pp. 501-516.

Rasmussen, J., Schmitz, H. and van Dijk, M. (1992), 'Introduction: exploring a new approach to small-scale industry', *IDS Bulletin*, vol. 23, no. 3, pp. 2-6.

Reinert, E. and V. Risser (1995), 'Recent trends in economic theory: implications for development geography', in J. Hesselberg (ed), *Development in the South: Issues and Debates*, Department of Human Geography, University of Oslo, Oslo, pp. 9-33.

Schmitz, H. (1993), *Small Shoemakers and Fordist Giants: Tale of a Supercluster*, IDS Discussion Paper No. 331, IDS, Sussex.

Schmitz, H. (1998), *Responding to Global Competitive Pressure: Local Co-operation and upgrading in the Sinos Valley, Brazil*, IDS Working Paper No. 82, IDS, Sussex

Schmitz, H. and Musyck, B. (1993), *Industrial Districts in Europe: Policy Lessons for Developing Countries?*, IDS Discussion Paper No.324, IDS, Sussex.

Sengenberger, W. and Pyke, F. (1992), 'Industrial districts and local economic regeneration: Research and policy issues', in F. Pyke and W. Sengenberger (eds), *Industrial Districts and Local Economic Regeneration*, Institute for Labour Studies, Geneva.

Shohat, E. and Stam, R. (1994), *Unthinking Eurocentrism: Multiculturalism and the Media*, London, Routledge.

Slater, D. (1989), 'Peripheral capitalism and the regional problematic', in R. Peet and N. Thrift (eds), *New Models in Geography, Volume 2*, London, Unwin Hyman, pp. 267-294.

Slater, D. (1992), 'On the borders of social theory: learning from other regions', *Environment and Planning D: Society and Space*, vol. 10, pp. 307-327.

Smelser, N. and Swedberg, R. (eds) (1994), *Handbook of Economic Sociology*, Princeton, Princeton University Press.

Stein, H. and Wilson, E. (1993), 'The political economy of Robert Bates: A critical reading of rational choice in Africa', *World Development*, vol. 21, no. 6, pp. 1035-1053.

Sverrisson, A. (1992), 'Flexible specialization and woodworking enterprises in Kenya and Zimbabwe, *IDS Bulletin*, vol. 23, no. 3, pp. 28-34.

Tarrow, S. (1996), 'Making social science work across space and time: A critical reflection on Robert Putnam's *Making Democracy Work*', *American Political Science Review*, vol. 90, no. 2, pp. 389-397.

Thomas, J.J. (1995), *Surviving in the City: The Urban Informal Sector in Latin America*, Pluto Press, London.

Thrift, N. and Olds, K. (1996), 'Refiguring the economic in economic geography', *Progress in Human Geography*, vol. 20, no. 3, pp. 311-337.

Toye, J. (1993), *Dilemmas of Development*, Blackwell, Oxford.

van Dijk, M. (1992), 'How relevant is flexible specialisation in Burkina Faso's informal sector and the formal manufacturing sector',. *IDS Bulletin*, vol. 23, no. 3, pp. 45-50.

Waldinger, R. (1995), 'The "other side" of embeddedness: a case-study of the interplay of economy and ethnicity', *Ethnic and Racial Studies*, vol. 18, no. 3, pp. 555-580.

White, G. (1993), 'The political analysis of markets: Editorial introduction', *IDS Bulletin*, vol. 24, no. 3, pp. 1-11.

Williams, D. (1999), 'Constructing the economic space: The World Bank and the making of *homo oeconomicus*', *Millennium: Journal of International Studies*, vol. 28, no. 1, pp. 79-99.

Williams, D. and Young, T. (1994), 'Governance, the World Bank and liberal theory', *Political Studies*, vol. 42, pp. 84-100.

Williamson, O. (1985), *The Economic Institutions of Capitalism*, Free Press, New York.

Ypeij, A. (1998), 'Transferring risks, microproduction, and sub-contracting in the footwear and garment industries of Lima, Peru', *Latin American Perspectives*, vol. 25, no. 2, pp. 84-104.
Zukin, S. and Di Maggio, P. (eds) (1993), *Structures of Capital: The Social Organization of the Economy*, Cambridge University Press, Cambridge.

Chapter 4

Supply Chains, Embeddedness and the Restructuring of Argentina's Tanning Industry

Kjersti Wølneberg

Introduction

Current views on successful local economic development emphasise the local embeddedness and integration of business within strong socio-economic networks as a source of local social capital that, in the right circumstances, can create self-sustaining growth. That growth, in addition, is said to be robust in an internationally competitive environment that is constantly being reinforced by processes of economic globalisation. This interpretation of successful local growth, built on institutionalist foundations, has been most extensively and elaborately promulgated in the developed country context. The interpretation also raises the Third Italy to the status of a type example, with a range of hi-tech centres in the US achieving similar status (Braczyk *et al.* 1998, Clark *et al.* 2000, OECD 2001). However, despite the World Bank's enthusiasm for this model as a basis for charting and planning a path towards economic development in the countries of the Third World (World Bank 1997, Fine 1999) empirical studies of economic clustering in these countries tell a very different story. Self-reinforcing local economic growth appears rarely to be achieved (Taylor 2001a). Instead, multinational capital, buyer-driven supply chains and apparently inappropriate governance structures are seen to deflect local economies away from realising their full and independent growth potentials.

The purpose of this chapter is to explore the interplay of multinational and local capital, national and international governance structures, and the dynamics of global supply chains in shaping the prospects of Argentina's tanning industry through the 1980s and 1990s. Through this analysis it is possible to begin to develop an appreciation of the economic and commercial processes and practices that limit economic success in this 'second world' context (*The Economist* 2000). It also provides a way of reflecting on the role, nature and prominence of embedded inter-firm relationships and local social capital in shaping local economic fortunes.

Argentina is a major world producer of hides, putting an estimated 16 million on to the domestic and international markets annually (Wølneberg 2000). The tanning industry is principally located in the industrial suburbs of Buenos Aires where currently 150 companies operate (Wølneberg 2000). In the past 20 years, the national and international environment within which Argentina's tanning industry operates has changed radically:

- the structure of international production has changed organisationally and locationally along with the international fortunes of buyers and suppliers;
- the technology of production has changed, especially in the chemical processing of hides;
- the international regulatory regime faced by companies in the industry has simultaneously become in some places more hostile and in others more amenable; and
- Argentina's national politics have shifted from dictatorship that fostered protectionism to democracy as a champion of deregulation.

The interplay of these elements of political economy during the past 20 years has created a distinctive set of attitudes and relationships among the companies of Argentina's tanning industry, and between those companies and international buyers and suppliers. The question addressed in this chapter is whether these attitudes and relationships constitute a mechanism for the creation of social capital that might foster self-sustaining local growth in this developing country context. The primary and secondary data on which the study is based were collected in Buenos Aires between September and December 1998. In-depth interviews have been used to elaborate the nature and form of inter-firm relationships and to reflect on the extent of local embeddedness[1].

The argument of the chapter is developed through three main sections. In Section 1, the mechanisms of 'knowledge transfer', 'learning' and innovation that might generate local social capital are outlined briefly in the 'clustering' and 'commodity chain' contexts. In Section 2, the massive shifts in the domestic and international regulatory regimes and in the regime of international competition that have confronted Argentina's tanning industry are explored. In Section 3, the nature of current inter-firm relationships at both local and international scales is examined to reflect on the existence of mechanisms of 'learning' that might create local social capital. From this analysis of the tanning industry it is concluded that the machinations of international capital and their interplay with the local governance regime in Argentina at best restrict and at worst inhibit processes that might create local social capital and the self-sustaining local economic growth that Argentina seeks.

Learning and Social Capital in 'Clusters' and 'Commodity Chains'

As globalisation deepens, a major question concerns how places in both developed and developing countries will be incorporated into this emerging global economic system. Current thinking envisages 'success' at the local level now being built through increased interdependence between firms and business enterprises and their incorporation into networks of trust-based relationships. The networks themselves are envisaged as assuming two distinctive but interrelated forms. On the one hand, they might form strongly territorial local 'clusters' – agglomerations, industrial districts, innovative milieu or 'learning' regions (Asheim 1997, 2000). On the other hand, they might form more geographically spread but just as functionally tight-knit commodity chains or supply chains (Gereffi 1994, 1999, Knutsen 2000). It is quite conceivable, however, that individual firms might be members of both types of network, especially when the 'clusters' within which they are embedded are export orientated, as has been seen to be the case in East Asia, India, the Caribbean, Brazil and Argentina, for example.

Indeed, these two types of networks are seen as being strongly complementary. In both circumstances, the engine of success is identified in processes of 'learning' and the mobilisation of knowledge through socially constructed, trust-based, inter-firm relationships (Schmitz 1999). However, while 'clustering' processes involve cooperation and collaboration rather than price-based competition, 'commodity chain' processes implicitly involve the benign exercise of power in conditions of, sometimes, gross inequality between buyers and suppliers. The processes are, nevertheless, reckoned to stimulate invention, innovation and technological change. These in turn, create and maintain high levels of productivity and competitive advantage within 'clusters', fostering enterprise upgrading and the competitive strength of firms in commodity chains. Fundamental to these interpretations of place-based 'success' in the face of globalisation are embedded *mechanisms of economic and social inclusion* based on trust, reciprocity and loyalty amongst firms, both SMEs and larger enterprises (Taylor 2001b). For Coleman (1988) these relationships of trust create *social capital*, which is seen as a 'resource for action' (p. 95). As it was explained by Coleman (1988, pp. 100-101);

> Social capital ... comes about through changes in the relations among persons that facilitate action. If physical capital is wholly tangible, being embodied in observable material form, and human capital is less tangible, being embodied in the skills and knowledge acquired by an individual, social capital is less tangible yet, for it exists in the *relations* among people. Just as physical capital and human capital facilitate productive activity, social capital does as well. For example, a group within which there is extensive trustworthiness and extensive trust is able to accomplish much more than a comparable group without that trustworthiness and trust.

The question addressed in this chapter is whether these postulated relationships and the resultant social capital currently exist in the concrete empirical circumstances of the apparent 'cluster' of tanning enterprises in Buenos Aires in Argentina. Two questions are addressed in parallel with this question: how are inter-firm relationships in this industry affected by shifting power relations in the global commodity chain of which the tanning industry is part; and, how has deregulation and liberalisation of Argentina's economy affected inter-firm relationships?

Argentina's Tanning Industry and Changing Regimes of Regulation

Protection and Local Growth: the 1970s and 1980s

In the early 1970s, virtually all the hides from the 16 million cattle slaughtered annually in Argentina were exported in an initially processed state as salted hides or 'wet-blues'[2]. The government of the time wanted to encourage more value adding in the country and so enacted in 1972 measures that effectively banned exports of salted and wet-blue hides. As a result, the number of tanning companies in Argentina grew to some 300 in 1989, of which some were very large while others were SMEs. What developed in Argentina, at just the time that the international market for leather was globalising, was a comfortable, non-competitive commercial regime, hidden behind barriers of protection. There were no incentives and no will to improve levels of productivity. The industry was technologically backward. It was easier to make profits from trading currencies on the black exchange markets than from trading products.

The tanning industry that developed was strongly localised in the industrial suburbs of Buenos Aires, principally Lanús and Avellanedua, which at one time housed more than 200 tanneries (now only 80). This concentration of tanners was associated with a number of leather-using industries, especially shoe making, and would outwardly appear to have constituted an industrial 'cluster' in the functional sense used by Porter (1990, 1998). If it was a cluster, however, it owed its existence to government regulation and protection.

But, the protected regime within which Argentina's tanning industry worked in the 1970s and 1980s came under pressure from three different directions at the beginning of the 1990s, from:

- changes in the nature and functioning of the buyer-driven international commodity chain dealing in leather;
- retaliatory duties imposed by the US on Argentina's exports to the US and continuing tariff barriers erected by the EU; and
- hyperinflation in Argentina.

These pressures brought a shift in Argentina's regulatory regime, involving the first steps towards deregulation and trade liberalisation, and a repositioning of Argentina in global commodity chains trading in leather.

Deregulation and Trade Liberalisation

At the end of the 1980s, Argentina's economy experienced severe instability and hyperinflation. The presidential elections of 1989 brought Carlos Menem and his Justicalist party[3] to power. The Menem administration's answer to the economic crisis was an integrated set of neo-liberal policies of deregulation and trade liberalisation in line with the recommendations of the IMF and the World Bank. At the heart of the reforms was the *Convertibility Law* ('convertibility'), which came into force on 1 April 1991 (Neffa 1998, Carassai 1998). The law established exchange parity between the Argentinean peso and the US dollar, and made domestic money supply directly dependent on the level of the country's foreign currency reserves.

From 1991, Argentina's economy was also substantially deregulated[4] (Neffa 1998, Russo 1998, *The Economist* 2000). The government eliminated price control, reduced the regulation of foreign investment and liberalised the exchange market. Customs formalities were made more pliable and national purchase laws were abolished. Deregulation, however, faltered as the Menem government became tainted with corruption and was replaced by the de la Rua government in December 1999 (*The Economist* 2000).

Trade liberalisation was an essential part of economic deregulation. The Menem government began to open the domestic market to imports by reducing tariffs, removing non-tariff barriers, and through 'convertibility'. In 1991, Brazil and Argentina negotiated a bilateral trade agreement for the products of their leather, automobile and sugar industries. Part of this deal, which opened Argentina's markets to Brazil, involved the phasing out of Brazil's tariffs on salted and wet-blue hides from Argentina, with the total removal of those tariffs by 2000. Central to Argentina's strategy of trade liberalisation was the reduction of the price of imported capital goods, to enable firms to increase their productivity (Neffa 1998). These measures brought about the restructuring of a number of tanneries, but the impact on exports was relatively limited and indirect. In 1994, however, the government decided to refund value-added tax (VAT) on exported industrial products. This was of vital importance to exporters, since the VAT rate was set at 21%.

However, Argentina's protectionist policies in the leather industries, and indeed in most sectors of the economy (*The Economist* 2000), have not been dismantled as fast or as effectively as some countries would have liked. Protectionism was condemned by US competitors with the result that the US government imposed a countervailing duty on leather from Argentina for a number of years during the 1990s. The duty hit Argentina's tanneries hard, and its lifting in 1997 caused a sharp rise in the export of Argentinean leather to the USA in subsequent years. The only remaining regulatory disadvantage for Argentina's tanners is a 7% import tax imposed by the EU on Argentinean crust and finished leather since 1997. As a result, Europe's share of Argentina's leather exports fell from 33% in 1994 to 17% in the first half of 1998 (Cuero No. 381 and 390).

The 1990s economic reforms had a major impact on the tanning industry by both enforcing and enabling restructuring. A manager of one of Argentina's biggest tanneries claimed that the tremendous productivity improvement seen in the country's tanning industry in the 1990s would not have been possible had it not been for the Convertibility Law (interview T10).

The main enforcing factor, however, was the rise of what was known locally as 'the Argentinean cost', which squeezed the profit margins of Argentina's tanners. The exporting tanners had been earning in dollars and paying expenses in the weak national currency. With 'convertibility', materials and wages had to be paid in dollars and a rise in the strength of the dollar meant more expensive production and reduced competitiveness. At interview, one manager reported that while a shop-floor worker earned approximately US$ 200 during the period of hyperinflation between 1989-1991, his wages increased to some US$ 600 with 'convertibility' (interview T2). The weakening of European and Asian currencies against the US dollar in the second half of the 1990s served to further restrict Argentina's exports of leather.

The pegging of the peso to the US dollar also removed the rationale for illegal dollar imports in Argentina, which were said to have been a lucrative business for some export tanners during the time of hyperinflation[5]. At times, the difference between the official exchange rate and the black market exchange rate had been as much as 40%. By under-invoicing exports of leather, some tanners had secretly brought in dollars and earned high profits on the black currency markets. The then economic reforms produced both higher production costs and marked the end of a profitable black currency market, reinforcing the already pressing need to rise productivity in the industry.

Argentina's tanners had long been aware of the need to upgrade their technology, to be internationally competitive, but lacked access to updated technology. As one interviewee remarked:

> Argentina didn't have access [to technology], because it was very complicated to import. ... There was no financing. The Argentinean machinery industry was strongly protected despite making very bad machinery. The tariffs were also very, very high. (interview T10)

Economic reform changed this situation drastically and became a crucial enabling factor in the restructuring of the tanning industry. The most concrete measure was the elimination of tariffs on imported capital goods, which greatly reduced the costs of investment in new technology. But, this opportunity could only be realised when 'convertibility' allowed foreign currency reserves to be reconstructed and credit to be made available.

'Convertibility' also tamed inflation and created a more stable and predictable climate, which was essential for corporate budgetary planning:

> Today we work with better budgets. This is very important. If we enter a contract with a customer based on a cost estimate, we know now that there is a certain stability, so that we won't suddenly lose money because of a change in costs. (interview T6)

For the big companies, then, liberalisation, deregulation and 'convertibility' created the necessary conditions for raising productivity. A reduction in the size of the workforce and improvements in product quality, as a result of using new technology, offset the rise in 'the Argentinean cost' that this brought. Simultaneously, this use of new technology also helped to position these bigger Argentinean tanners in the emerging international markets for upholstery and sport shoe leathers.

For smaller tanneries that mainly served the national market, however, the consequences of the liberalisation were severe, mainly because of its impact on their clients. The opening of Argentina to imports of cheap Brazilian and East Asian leather shoes caused a crisis in the national shoe industry. Shoe imports increased from 1.5 million pairs in 1990 to 15 million pairs in 1995 (Herzovich 1997). In the same period, 700 shoe producers closed down in Argentina (Herzovich 1996). For the tanners, this meant a severe contraction in the national market for footwear leather.

Changing Commodity Chain Relationships in the Leather Industry

In parallel with, and partly as a consequence of, changes in the national and international regulatory regime faced by Argentina's leather tanners, their positioning in the global commodity chains dealing in leather has changed substantially. While their commercial opportunities have increased, so too has their vulnerability.

The markets for leather are in footwear, upholstery, garments and leather accessories (bags, purses, belts etc.), and those markets have rapidly internationalised in the past 20 years (Ballance *et al.* 1993). Argentina's tanners specialise in the footwear and upholstery leather markets. Both markets are demanding, and increasingly buyers are requiring ISO certification to guarantee quality and reliable deliveries. However, demand conditions have changed very differently in these two markets in the last 20 years. The demand for medium priced upholstery leather has risen significantly (Ballance *et al.* 1993, interviews: O5, O6), and this has presented Argentinean tanners with the opportunity to upgrade their technology and to enter the growing North American market and the emerging markets of Malaysia and China (interview T2). In contrast, the producers of shoe leather have experienced harder times. Shoe production has shifted from Europe and the former Soviet Union to East Asia, where competition from upcoming Asian tanneries is increasing. More recently, demand for shoe leather in Argentina has also fallen as a result of economic liberalisation and the opening of the domestic market to imports. The other lesser markets for Argentina's leather, for garment and accessories leathers, have been either stable or declining. Exacerbating these problems, international prices in the lower quality market segments have been squeezed. This has been to Argentina's disadvantage because it is a higher cost producer than other leather producing developing countries.

As part of these changes in demand, the international map of leather production has also changed along with the functional division of labour between countries. Two major shifts in location can be identified. First, there has been a shift of production from high-cost to low-cost countries within Europe. From the late 1960s, traditional tanning countries such as Germany, France and England experienced recession, while the Italian and later the Portuguese and Spanish industries grew. Second, and subsequently, low-cost developing countries of the South have gained market shares at the expense of the traditional tanning countries of the North.

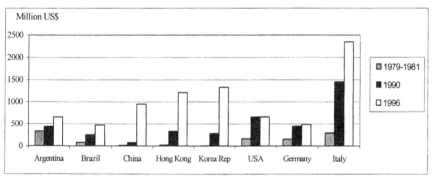

Source: FAO 1998.

Figure 4.1 Leading leather exporters, 1979-1981, 1990 and 1996

Figure 4.1 shows the export performance of the world's main leather exporters in the last 20 years. The surge of exports from Asian countries, especially China, Hong Kong and the Korean Republic is only too clear. India is a fourth important Asian tanning country, but exports little. Hong Kong, however, does not produce leather but has become the main trade centre for leather in Asia.

This success in Asia is closely tied to the growth of leather shoe production, and is contributing to the declining leather production in Europe and the USA. However, while volumetric leather production is declining in the USA and Germany the value of their exports is rising – a consequence of their specialisation in the production of high quality finished leathers. Italy, however, is the principal leather tanning country in the world, and its position both as producer and exporter has been confirmed by a massive growth of exports in the past 20 years.

Behind these figures lies a clear division of labour that parallels the North-South divide (Knutsen 1998). The remaining tanners of the North cater to the high quality markets, while the tanners in the South, including Argentina, mainly serve the lower quality segments. There have also been profound changes in the organisation of the leather production chain during the last decade. Knutsen (1998) has identified a new intra-sectoral technical division of labour that involves increased process specialisation in the leather industry, at both firm and national levels (Gjerdåker 1999). Most tanneries now specialise in one market segment, such as upholstery or footwear leathers. Most firms specialising in upholstery leathers are located in the North but, during the 1990s, some countries in the South, such as Argentina, entered the market. The territorial dimensions of specialisation are even more striking. Tanners in low-cost countries are to an increasing extent becoming suppliers of 'wet-blues' (initially processed leather) and sometimes 'crust' (semi-processed leather) to tanners in high-cost countries, who concentrate on the more rewarding and less polluting finishing processes. This shift has occurred because of the crisis faced by tanners in many low-cost countries, such as Brazil and Poland for example, brought on by the decline of their national footwear industries. The liberalisation of these economies that has been adopted as a panacea for crisis has also allowed foreign investors to enter these exposed industries.

Argentina has not experienced the same process of degrading of its leather tanning industry. However, as a supplier of medium-priced leather it faces tough competition as companies in high-cost countries seek sources of cheap wet-blue hides from abroad. Internationalisation in the tanning industry has endowed the bigger and financially stronger companies with the power to reduce their production costs by adopting aggressive sourcing strategies.

The Argentinian Tanning Industry in the 1990s

Figures on Argentina's leather exports in the 1990s paint the picture of a competitive, growing industry. Exports of leather[6] increased by 69% between 1991 and 1997, from 482 million US$ in 1991 to 816 millions in 1997[7]. These exports were principally of finished leather, and exports of crust and finished leather from Argentina supplied 9% of the world market in 1997 (interview O6). The improved prices that were obtained were a consequence of massive investment in new technology and the improved quality this yielded.

A sign of the major changes in the industry is the shift in the main export markets for Argentinean leather. Figure 4.2 depicts the developments in the seven most important export market countries during the 1990s.

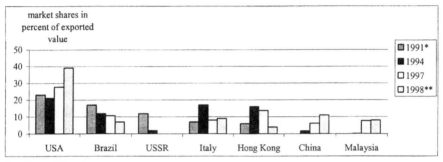

*Jan-Sep, **Jan-Jun
Source: CICA 1991, Cuero No. 381, 389 and 390.

Figure 4.2 Leather exports from Argentina, 1991-1998, by country

The USA has long been an important market for Argentinean upholstery leather, and the biggest importer of the country's leather in the late 1990s. Other traditionally big markets, in Brazil and the former Soviet Union have, respectively, declined and disappeared. Leather footwear production has shifted from the former Soviet Union to Asia. Since the beginning of the 1990s, Hong Kong has become a major export market, and recently China and Malaysia have become significant importers of Argentinean leather[8].

Despite these positive export figures, most tanners in Argentina view the 1990s as a period of crisis. They did not participate in the global export boom and, in the period 1988 to 1998, the total number of tanners in Argentina declined from some 300 to about 150 (interview O6). Closures and bankruptcies haunted the industry, hitting both big and small companies. Restructuring favoured only a few innovative companies while weaker companies were squeezed out. In the first half of 1998, the three largest tanneries (who are among the world's biggest) accounted for 33% of the value of exported leather, and the 12 largest for 84% (Cuero No. 390). The question arises, therefore, as to why the fortunes of Argentina's tanners should be so perverse in an international environment of expanding demand. The answer appears to lie in the nature of international commodity chain relationships rather than local inter-firm collaboration, and this issue will be explored in the next section of the chapter.

Doing Business in the Commodity Chain

In a shifting international economic environment what strategies can firms adopt to respond to change? Applying the structure of the commodity chain, I will now have a closer look at how contextual factors and firm strategies together explain the competitiveness of the industry. The evidence from Argentina's tanning companies suggests that local strategies are not the answer, as the cluster model of embedded local growth would suggest. Instead, local companies are seen as relatively powerless when confronted with the international buyer-driven relationships of the commodity chain that tend to favour large companies. This powerlessness is only too clear in the supplier, production technology and customer links of Argentina's tanning companies.

Supplier Links

Empirical evidence demonstrates clearly that the exercise of power rather than collaborative trust is central to Argentinean tanners' transactions for inputs of hides. Hides make up between 50 and 70% of tanners' production costs (Knutsen 1999). Their scarcity on the international market, and the quality of Argentinean hides, brings fierce international competition for this input. Also, fluctuating prices make it important to be able to buy big when prices are low. Both these factors favour big companies with strong finances and an international scope.

Exclusive access to the country's rawhides was for a long time a competitive advantage enjoyed by Argentina's tanners, and it is claimed to have been central to the successful restructuring of the industry in the 1990s (interviews T2, O2, O5, O6). However, the late 1990s were a watershed for rawhide procurement as a result of economic reform. The gradual removal of export tariffs on wet-blue hides changed the power relations between the slaughterhouses and the tanners. The relationship had long been tense, and the slaughterhouses argued that they did not get the right price for their rawhides. Deregulation allowed slaughterhouses to export wet-blue hides (including high quality splits), or to use contract tanners and export crust, significantly increasing their returns. In 1999, many of Argentina's big slaughterhouses entered into collaborative business with big tanneries and strong international customers. Such collaboration was, for example, established between the giant slaughter and supermarket chain, Coto, the Italian owned tannery, Curtarsa, and the Seaton Trading Company[9] from the USA. This type of powerful alliance raises questions about Argentina's future role in the international division of labour in the tanning industry because they run counter to strategies of the 1990s that attempted to promote local value-adding. Coto and Curtarsa have no particular interest in maximising value-added in Argentina. Coto is a slaughterer and Curtarsa is an Italian owned tannery that does not have a finishing plant in Argentina. In fact, Curtarsa is part of a major Italian enterprise, Italian Leather, which has two finishing plants in Italy. When a big supplier of rawhides such as Coto decides to enter this kind of collaboration, it reduces the availability of rawhides for Argentinean tanners who export finished leather. Size endowed it with power.

The reaction of Argentinean tanners was to take measures to secure their raw material base. As a result, five of the country's major tanners[10] joined forces to compete for Brazilian wet-blue hides and to control hide prices in the local market. Once again, size brought power.

Size is again important in securing supplies of chemicals. A system of 'temporary import' in Argentina allows bigger tanneries to import processing chemicals directly from the producers, tax-free, when they are producing leather for export. Smaller, non-exporting companies have to rely on local dealers for their chemical supplies and consequently, pay a considerably higher price.

Chemicals technology is at the heart of technological development in the tanning industry, so links to chemical suppliers are vital. Now, collaboration for product development is widespread between tanners and chemical suppliers. This serves their mutual interest, since tanners need to develop new products to stay competitive and chemical producers seek to widen their markets. The need for collaboration is heightened by the complexity of the chemical mixes that tanners use, and the secrecy that surrounds these mixes because these are their competitive edge. The complicated mix of chemical processes makes it necessary to adjust a new product to each customer:

> The chemical producers generally introduce something for the whole industry. If it interests you, you receive them and work a week with them in order to meet the special needs of the technicians in the tannery. (interview T2)

Collaboration favours the larger tanners simply because of the size of their demand and the security that comes with size.

What is clear from this analysis is that the proximity and collaboration that has been postulated as generating learning processes, clustering and local growth by writers such as Porter (1998) does not hold for the supplier links of Argentina's tanners. Very clearly, supplier links in the industry are moulded by firm size and power inequalities between firms at the international trading scale. It is these inequalities that are the essence of trading relationships, modified by national regulation in Argentina and the regulatory power of countries like the USA, working in the interests of their own corporate lobbyists.

New Technology and the Production Process

The restructuring that deregulation brought to Argentina's tanning industry brought with it investment in new technology, and it has been estimated that 700 million US$ was invested in new machinery between 1991 and 1997 (interview O6). Perversely, this process served to reinforce the strength of the country's largest tanning enterprises.

This investment in new technology had many effects. First, it brought with it significant specialisation by the largest tanners who placed a new strategic accent on upholstery leather, and withdrew from the shoe leather and accessories leather markets. In this way they were able to deepen their specialist competencies. Second, by acquiring new leather finishing technology these large firms were able to add more value to their products. Third, they reduced production costs, not only by cutting input costs and reducing their use of manual labour, but also by streamlining their logistics and sales administrations.

Fourth, the new investments notably improved *standard* high quality (interview O3). Fewer irregularities and production flaws also reduced costs and boosted prices. Sadesa put great emphasis on these issues:

> The tanning industry is coming out of a paradox – whether it is a craft or an industry. It depends on where you choose to stand. For us it is an industry, like doing chocolate or cars or electronic equipment. We believe strongly in process engineering. ... We are convinced that with strong engineering, we can convert a non-standard raw material to a standard product. What we want to sell to our customer is a standard. If we can work with our customer in such a way that he opens one of our cases and puts it into production without having to control, then it is a service that we in a way add to the product. (interview T10)

The growing need in export markets for ISO certification of quality has reinforced this advantage of large company size.

The restructuring of the small and medium sized tanning companies was bound to be more limited due to the huge costs of investing in new technology. Retrieving these costs is always a challenge especially for firms producing smaller volumes. Still, the labour efficiency resulting from the introduction of new machinery was of major importance for the competitiveness of some SMEs (interviews T3, T1). For many companies, however, the lack of resources prevented big investments in new technology, weakening their competitive positions.

The Customer Link

Interviews with tannery companies show very clearly that power is moving towards the buyer in the leather industry commodity chain (Gereffi 1994, Gjerdåker 1999). As it was put by one Argentinean tanner:

> The customer's demands are harder and harder. They are less permissive than in earlier years, permissive with respect to possible errors and difficulties inherent in the leather. ... As supply exceeds demand, you have to comply with the new demands in order to keep on producing. The business is governed by the customers. (interview T2)

Gjerdåker has pointed out that in an industry like tanning, being linked to demanding customers, particularly overseas customers, is vital to the building of confidence and trust. This is especially important in the tanning business, as the industry is infamous for cheating (interview O6).

Four large and fast-growing tanning companies that were surveyed put particular emphasis on the importance of service delivery and customer satisfaction in maintaining their buyer links. To this end, they offered reliable and just-in-time delivery, flexible payment systems, and quick and easy problem-solving services. Some of the companies also had a strategy of reducing the number of customers they served in order to offer a better service and build partnerships with their remaining clients. In short, they aimed to become embedded with their principal customers.

The depth of that embedding extends to strategic collaboration. Espósito is one Argentinean tannery that has developed close collaboration with some of its customers both in marketing and product development. The company now exhibits, for example, at overseas furniture shows on the stands of its US customers, and even sends its own staff. This allows the tannery to build its own brand image, while it provides the furniture maker with specialist technical support. Collaboration also extends to product development. Also in product development, close collaboration with customers can be highly beneficial. The same company, again, tailors special leather qualities to the needs of a major US furniture manufacturer. This provides the furniture manufacturer with exclusivity and guarantees the tanner's sales of the new product. As it was expressed at interview:

> With Thomas Will [the US furniture manufacturer] it is no "let's see what happens". No, they do a market study on the launching of a new product line. Thus, it is practically a closed business. Thomas Will is an innovator. They introduce new products in the American market, and they introduce them very well. (interview T2)

What is important from this analysis, however, is not that collaboration and trust is important in developing inter-firm relationships in the tanning industry but that, in the 'second world' context of Argentina, geographical proximity is not a necessary factor for creating that trust. This finding resonates with Reinert's (1993) criticism of Porter's cluster theory that demanding customers do not necessarily have to be national.

Conclusions

This case study of the Argentinean tanning industry shows clearly that in this 'second world' national context local economic 'clustering' offers no competitive advantages. Instead, two sets of processes mould economic success in this industry, the interplay of governance structures at the national and international scales, and the nature of international inter-firm relationships. This study demonstrates the inequalities and unequal power inherent in these relationships and structures. Importantly, these inequalities produce seemingly perverse outcomes. Thus, while domestic protectionism (in this case in the tanning industry) can create apparent enterprise clusters, protectionism in international markets (especially in the US and EU) can destroy them. Equally, while multilateral deregulation, such as that introduced in Argentina, can open international markets to a nation's firms, it can, simultaneously, destroy segments of that nation's enterprises because of the advantage it gives to big business. Furthermore, while international inter-firm buyer-supplier relationships can offer growth potentials in economies like Argentina's, the advantages they bestow on corporate 'bigness' both undermine and restrict the spill-over effects that might favour domestic SMEs.

In effect, what this case study shows is that in a semi-peripheral economy like Argentina's, governance structures and international commodity chain relationships constrain to the point of crippling local growth potentials and opportunities to create self-sustaining 'clusters' of production. It can be suggested that a downward spiral of dependence is created in countries like Argentina as control of the domestic regulatory regime shifts into the hands of international financial agencies like the World Bank and IMF, as protectionism in powerful markets like the EU limits trade, and as international commodity chain relationships subordinate suppliers in favour of buyers. The spiral of intensifying dependence favours big business, promotes concentration and creates structural rigidities. It all but precludes local collaborative growth based on small enterprise, knowledge exchange and learning. Power inequalities, it would appear, far outweigh collaboration in shaping economic growth.

Notes

1 A case study strategy was adopted based on in-depth interviews with key informants from the industry combined with secondary sources of information. General information on the industry was obtained through six interviews with people in support institutions, a consultant, and a journalist specialising in the industry. These respondents are referred to as O1 through O6 in the text. Other information was obtained from trade magazines, statistics and literature on the field. In-depth information was collected on the experiences of individual companies in different segments of the industry. Because the upholstery leather market is the fastest growing segment of the industry, contributing heavily to the increasing value of exports, interviews were conducted with four tanneries specialising in this market. In-depth interviews were also conducted with six other tanneries spanning the range of activities in the sector. The tanneries from which information was collected are referred to as T1 through T10 in the text.

2 Tanning is a process that transforms raw hides into a stable product, leather, that cannot rot. To make leather, hides go through a range of physical and chemical processes. Today chromium is the most important chemical in this process but vegetable tanning agents are also used. Hides are normally split horizontally into two pieces of leather, the lower 'split' and the upper 'full grain' (split and grain). The full grain is the hair side of the hide, from which the highest quality leather products are made. The split is the flesh-side. It does not have any smooth surface and the structure of the leather is looser. The split has traditionally been a second grade or waste product, but recently technology has been developed that converts it into attractive leather. Leather is traded at three different stages in process:

(1) *Wet-blue* is leather that has only gone through the first so-called wet processes. In this stage the leather is wet and with a light bluish colour and it is a standardised product that can be applied in a range of different end products. The production of wet-blue is the most polluting and least rewarding part of the tanning process.

(2) *Crust* is a more elaborated half-fabricate. The leather is then rabbeted and greased and later dyed in drums. Having been dyed, the leather is vacuum dried, stretched and softened. Some crust is applied directly in shoe production.

(3) *Finished leather* is leather ready to go into the production of the final product. The finishing processes are various chemical and mechanical processes carried out to give the leather the right colour, brilliance and relief. The finishing technology is becoming increasingly sophisticated. The different operations include, among others, hydraulic ironing, printing of patterns and the application of paint, varnish and resin.

3   'Partido Justicalista' also called the Peronist Party.

4   Carrasai (1998) argues that 're-regulation' is a better concept to depict the changes. The 'deregulation of markets', he claims, does not do more than transferring the power of regulating the markets in question to particular economic agents.

5   This information was confirmed independently by four informants who preferred to remain anonymous.

6   Figures refer to 'light leather' produced from cattle only (i.e. not 'heavy leather' and leather from sheep and goat skins). 'Heavy leather' is associated with vegetable tanned leather for shoe soles, while light leather is primarily associated with chrome-tanned leather.

7   The statistics refer to current prices. Inflation has been low in the 1990s owing to the Convertibility Law of 1991, which pegged the Argentinean peso at parity with the US dollar.

8   The sudden drop in sales to Hong Kong in 1998 is explained by the Asian economic crisis. Hong Kong is the location of the main leather dealers in the Far East who keep enormous stocks from which they sell to customers all over East Asia. With the instability brought by the financial crisis, Hong Kong based companies chose to sell from their stocks of leather and wait for the situation to change (interview O6).

9   Seaton is a huge tanner that sources crust in Argentina, which the company finishes and distributes to customers such as General Motors.

10   Fonseca, Espósito, Becas, CueroArt and Curtarsa.

References

Asheim, B. (1997), '"Learning regions" in a globalised world economy: Towards a new competitive advantage of industrial districts?', in M. Taylor and S. Conti (eds), *Interdependent and Uneven Development, Global-Local Perspectives*, Ashgate, Aldershot, pp. 143-176.

Asheim, B. (2000), 'Industrial districts: The contribution of Marshall and beyond', in G. Clark, M. Feldman and M. Gertler (eds), *The Oxford Handbook of Economic Geography*, Oxford University Press, Oxford, pp. 413-431.

Ballance, R., Ghislain, R. and Forstner, H. (1993), *The World's Leather and Leather Products Industry*, UNIDO and Shoe Trades Publishing Ltd, Liverpool.

Braczyk, H-J., Cooke, P., and Heidenreich, M. (eds) (1998), *Regional Innovation Systems*, UCL Press, London and Bristol.

Carassai, S (1998), 'Un Mercado Que Elige, Un Estado Que Adbica, Un Crecimiento Que Posterga. La Experiencia Argentina de los Años '90', in H. Nochteff (ed), *La Economia Argentina a Fin de Siglo: Fragmentacion Presente y Desarrollo Ausente*, Edueba and Flasco, Editorial Universitaria de Buenos Aires, Sociedad de Economía Mixta, Buenos Aires.

CICA (1991), 'Brochure on the Argentinean Tanning Industry', Cámera de la Industria Curtidora Argentina, Buenos Aires.

Clark, G., Feldman, M. and Gertler, M. (eds) (2000), *The Oxford Handbook of Economic Geography*, Oxford University Press, Oxford.

Coleman, J. (1988), 'Social capital and the creation of human capital', *American Journal of Sociology*, vol. 94 (supplement), pp. 95-120.

Cuero No. 381, Agosto/Noviembre 1995, Publicación técnica especializada pr la Cámera de la Industria Curtidora Argentina, Buenos Aires.

Cuero No. 389, Abril/Julio 1998, Publicación técnica especializada pr la Cámera de la Industria Curtidora Argentina, Buenos Aires.

Cuero No. 390, Setiembre/Diceiembre 1998, Publicación técnica especializada pr la Cámera de la Industria Curtidora Argentina, Buenos Aires.

FAO (1998), *World Statistical Compendium for Hides and Skins, Leather and Footwear 1979-1997*, Food and Agriculture Organization of the United Nations, Rome.

Fine, B. (1999), 'The developmental state is dead – long live social capital?', *Development and Change*, vol. 30, pp. 1-19.

Gereffi, G. (1994), 'The organization of buyer-driven global commodity chains: How US retailers shape overseas production networks', in G. Gereffi and M. Korzeniewicz (eds), *Commodity Chains and Global Capitalism*, Praeger, Westport, CT, pp. 67-92.

Gereffi, G. (1999), 'International trade and industrial upgrading in the apparel commodity chain', *Journal of International Economics*, vol. 48, pp. 37-70.

Gjerdåker, A. (1999), 'Leather tanning in Scandinavia', *FIL Working Papers*, No. 18, Department of Sociology and Human Geography, University of Oslo.

Herzovich, M. (1996), 'Crecer hacia afuera', *Curtido y Calzado. La Revista Latinamericana*, September, pp. 37-41.

Herzovich, M. (1997), 'Cueros y manufacturas, dos escenarios diferentes, *Curtido y Calzado. La Revista Latinamericana*, September, pp. 149-162.

Knutsen, H.M. (1998), 'Restructuring, stricter environmental requirements and competitiveness in the German, Italian and Portuguese tanning industry', *Norsk Geografiska Tidsskrift*, vol. 32, pp. 167-180.

Knutsen, H.M. (1999), 'Leather tanning, environmental regulations and competitiveness in Europe: A comparative study of Germany, Italy and Portugal', *FIL Working Papers*, No. 17, Department of Sociology and Human Geography, University of Oslo.

Knutsen, H.M. (2000), 'Environmental practice in the commodity chain: The dyestuff and tanning industries compared', *Review of International Political Economy*, vol. 7, no. 2, pp. 254-288.

Neffa, J.C. (1998), *Modos de Regulación, Regimenes de Acumulación y sus Crisis en Ergentina (1880-1996)*, Asociacón de Trabajo y Sociedad, Buenos Aires.

OECD (2001), *Innovative Cluster: Drivers of National Innovation Systems*, OECD, Paris.

Porter, M. (1990), *The Competitive Advantage of Nations*, Macmillan, London.

Porter, M. (1998), *On Compeitition*, Harvard Business School Publishing, Boston, MA.

Reinert, E.S. (1993), 'Porter-prosjektet, økonomisk teori og fremtidig industripolitikk', *Fretek-Notat 1/93*, Norges Teknisk-Naturvitenskapelige Forskningsråd, Oslo.

Russo, C. (1998), 'La Argentina de los '90: Transformaciones macroeconómicas y reestructuración productiva', in H. Nochteff (ed), *La Economía Argentina a Fin de Siglo: Fragmentacion Presente y Desarrollo Ausente*, Edueba and Flasco, Editorial Universitaria de Buenos Aires, Sociedad de Economía Mixta, Buenos Aires.

Schmitz, H. (1999), 'From ascribed to earned trust in exporting clusters', *Journal of International Economics*, vol. 48, pp. 139-150.

Taylor, M. (2001a in press), 'Enterprise, embeddedness and exclusion: business and development in Fiji', *Tijdschrift voor Economische en Sociale Geografie*.

Taylor, M. (2001b), 'Enterprise, embeddedness and local growth: inclusion, exclusion and social capital', in D. Felsenstein and M. Taylor (eds), *Promoting Local Growth: Process, Practice and Policy*, Ashgate, Aldershot, pp. 11-28.

The Economist (2000), 'A survey of Argentina: Getting from here to there', *The Economist*, May 6, 16 pp.

World Bank (1997), *The State in a Changing World* (World Development Report), Washington DC, World Bank.

Wølneberg, K. (2000), *Competitive Leather: A Case Study of the Argentinean Tanning Industry in the 1990s*, unpublished thesis for the Cand. Polit Degree in Human Geography, University of Oslo.

Chapter 5

Going Places? Reflections on Embedding and Disembedding in Agriculture and Horticulture Under Neoliberalism: The Example of Hawke's Bay, New Zealand

David Hayward, Christina Stringer and Richard Le Heron

Introduction

New Zealand's agricultural industries offer a unique set of conditions through which the dynamics of embedding and disembedding may be examined. These are formed by the conjuncture of state-led industrial restructuring, and the responses of New Zealand's global food commodity chain corporates and major trading partners to world economic conditions over the past decade (Le Heron and Pawson 1996). This chapter explores embedding and disembedding in the context of the reconstitution of three distinctive 'logics' of industrial practice and performance associated with the meat, apple and vegetable industries in Hawke's Bay, New Zealand during the 1990s. Rather than begin with an a priori definition of 'embeddedness' or launch into an extended discussion of theoretical perspectives on the phenomenon, we take a different approach. First, we briefly construct the mix of economic and institutional processes that make up and shape the regional components of the three commodity chains under study. Second, we use this to sketch the possibilities and constraints inherent in the evolving structural relations of each chain. This forms the basis of an exploration of the changing geographic scale of embedding and disembedding tendencies associated with these chains. Through this approach we present an analysis that questions Gereffi's (1999) recent conclusion that organisational learning and industrial upgrading follow from firms participating in global commodity chains. We suggest that the three Hawke's Bay chains show different sets of responses over time and show no real convergence with Gereffi's upgrading hypothesis.

The embeddedness concept introduced in Chapter 1 implies a local spatiality to the process of embedding. Following this orthodoxy we take a 'region' in New Zealand as a point of entry. From the perspective of New Zealand in the mid-1980s, Hawke's Bay would have been regarded as a thriving, diversified, export-oriented regional economy centred on land-based production encompassing agriculture, horticulture, forestry and viticulture. However, Hawke's Bay has experienced changing economic fortunes with peculiar local features brought about by its specific context. For instance, prior to the 1980s the region's economic base had benefited from diversification through state fostered industrialisation (Le Heron 1980) but subsequent neo-liberal political policies removed this factor. A further appeal of Hawke's Bay as a study region is the growth potential at that time of the region's two medium-sized cities – Hastings and Napier – through the growth-inducing effects of inter- and intra-industry linkages involving primary resource processing. Other distinctive structural features of the local economy include the region's asymmetrical enterprise structure, consisting of a small number of agri-processing firms drawing on inputs from a large number of farmers and growers. Changes in the structural context of the three study chains posed a severe test of the ability of regional producers to weather the evolving operational and institutional environments in which they were set. At issue is whether embedding and disembedding processes influenced regional economic outcomes and fortunes within global commodity chains.

The chapter reflects upon the findings of a series of projects undertaken in the region in the 1990s that considered transitions in land-based production. A common thread in this research has been a longitudinal assessment of regional growth processes and the spatiality of economic change. Events in the late 1990s reveal both dimensions of stability and instability in industry and regional processes. It is a methodological necessity, therefore, to conceptualise and investigate the spatiality of embedding and disembedding in the three study chains from a dynamic perspective.

Based on earlier work (Cooper *et al.* 1995, Hayward *et al.* 1998) our premise is that the land-based industries in Hawke's Bay are best considered as a mix of co-existing, localised components of commodity chains. In a sense, this interpretation is counter-intuitive in two respects. First, the visible rural and urban economic landscape is suggestive of an agri-industrial district, featuring strong and supportive linkages; although this view overlooks the connections of the region to the larger, globalising economy (Markusen 1999). Second, the embeddedness literature exhibits an evolutionary flavour; it is usually developed as a linear sequence from proximity, specialised area, industrial district, learning district to innovative milieu (Capello 1999, Hwang and Choo 2000). Indeed, Gereffi (1999), with respect to the apparel industry, proposes the upgrading hypothesis with much in common to that found in the learning region's literature (e.g. Storper 1992). The lengthy and competitive history of meat, apple and vegetable production in the region could suggest the conclusion that the region has acquired special attributes – or social capital, perhaps – giving it a global competitiveness. However, we take an open position on both the embedding and disembedding realities associated with the region, and the most appropriate label that might be assigned to describe the economic and institutional patterns and the geography of interactions and supportive structures. Most importantly, we resist ascribing the horizontal linkage typology developed for manufacturing industries (e.g. industrial districts or learning regions) or its latter-day commodity chain variant espoused by Gereffi (1999), although we believe that there may be points of intersection in theory which might be informed by our agro-industrial analysis.

Co-existing Industrial Logics in Hawke's Bay

This section introduces the three regional industries and briefly establishes the trajectories of industrial change in the separate food commodity chains. The significance of the regional context and the limited development of interdependencies highlight the constraints on the embeddedness model while at the same time illustrating embeddedness processes. The review of the industries reveals very different spatial processes as part of the evolving context of enterprise and industry.

The Meat Industry: From Spatial Concentration to Regional De-industrialisation

New Zealand's meat industry is a long-established leading export sector but has undergone massive restructuring since the mid-1980s. This has been the result of pressures from at least three directions: a reduction in livestock numbers following the removal of agricultural assistance since 1984; a resulting over-capacity and intense competition among the large-scale, second-generation processing-freezing works; and the entry of small-scale, third-generation works following industry de-licensing in 1981 (Curtis

1999, Le Heron 1991a, 1992, Lynch 1996). One of the first closures in the national rounds of restructuring was Hawke's Bay's Whakatu works in 1986; the largest and most modern second-generation works in New Zealand. The mechanism by which industry rationalisation was effected on this occasion is pertinent. The choice of which meat works to close was decided by a company especially created by the major North Island companies to manage a reduction in industry capacity[1]. Regionally, however, the closure was absorbed because of the presence of another large-scale plant (Weddell's Tomoana works) and several innovative, small-scale meat processors: Hasting Beef Processors, Hill Country Beef, Lowe Walker, and Richmond. Subsequently, the Weddell company began to attract attention as a progressive processor, embarking in the early 1990s on a total quality management (TQM) programme and supporting broader research and development in the industry (Perry *et al.* 1995). Meanwhile the third-generation companies effected a specialised division of labour with local connections focusing on stock procurement contracts and marketing-distribution (Stringer 1999). However, what was described by Perry *et al.* (1995) as a regional commitment to skills enhancement on the part of Weddell's work practice changes, was unexpectedly terminated in 1995 when its British parent company (Vesty International) ceased its New Zealand operations; a decision made as part of an international debt-settlement package and owing little to the Hawke's Bay industry. In the six years since the Weddell-Tomoana closure, Hill Country Beef has also closed, Lowe Walker has been merged into Richmond (with a local plant closure) and, in turn, Richmond has been acquired by a diversified corporate raider – the Brierley Group (see Le Heron (1991b) for a discussion of the Brierley Group and its strategies).

The case of the Hawke's Bay meat industry is instructive in terms of the proximity thesis in regional growth theory, and the complementary notions of structural embeddedness (Taylor 1998, Taylor and Leonard this volume). The industry was largely developed under a peculiar regime of state agricultural support. Specifically, favourable cost-plus conditions guaranteed producers' incomes through schemes such as the Supplementary Minimum Payments in the period from the late 1970s to the early 1980s (see Le Heron 1989a, 1989b). In this context, the national meat industry comprised a series of relatively independent and often isolated plants, and Hawke's Bay was no exception being host to seven processing works in the mid-1980s. The advantages of proximity in the conventional sense were largely illusory and interdependence was not evident. Each meat works operated independently and buyer competition for livestock was often vigorous. Nationally, meat works investment was influenced more by European Union hygiene regulations than product and processing innovation.

In retrospect, a number of issues might have been aired regionally had a different industrial culture prevailed. Some relevant questions are: What regional learning might have been pursued (and by whom?) that might have averted the big company closures, or given better prospects for the speciality meatworks? Or could the works management at, for instance, either Whakatu or Tomoana have channelled sufficient regional political punch within the corporation to postpone or prevent the closures? Perhaps these are simply the questions of academic optimists, placing too much faith in the power of local agency to influence the nature of local outcomes in multi-regional companies? The scarce evidence suggests that the regional components of the different companies did not collaborate to establish a local processing culture beyond cooperation over effluent disposal and common infrastructure cost (Hayward *et al.* 1996, Stringer 1999). The industry dynamics can thus be seen as an initial spatial concentration in the region followed by a phase of decline culminating in a significant loss to the region's economic base. Perhaps the key conclusion here is that the absence of strong local 'associations' reflected structural forces within the industry; that is, individual company strategies being entrenched in a logic of competitive independence with respect to both inputs and outputs. Furthermore, had they existed, local interdependencies would have made little difference in the corporate forum that sealed the fate of Whakatu or the international debt restructuring by Vesty.

Apple Industry: Regional Specialisation Under a Producer-Marketing Board Framework

In contrast to the meat sector, the development of the apple industry over the past 50 years has been under the regulatory regime of a statutory producer-marketing board; a specific model of market governance (Christopherson 1999). A production environment sympathetic to growers was ensured by the cooperative foundation of the Apple and Pear Marketing Board (APMB). The APMB took overall responsibility for the production chain as well as the initiative for chain development, fruit packing, fruit quality, research and marketing (Le Heron and Roche 1996, Peterson and Holland 1996). Growers accepted the monopoly position of the Board as exporter and domestic seller of all New Zealand fruit, as the concomitant APMB obligation to accept all fruit presented to it engendered certainty and stability for growers (Blunden 1996). In Hawke's Bay in the mid-1990s approximately 500 growers and 100 packhouses supplied the APMB through its marketing company, ENZA (Cooper *et al.* 1995).

Table 5.1 Industry perceptions of the Apple and Pear Marketing Board's performance

Key advantages

- Provides reliable coordination of the cool chain

- Provides packhouse capacity as required

- Coordinates and ensures compliance of hygiene and pesticide regulations

- Development of long term market relations as a single-desk seller

- Able to capture price premiums on new varieties for industry as a whole

- Maximised long-run returns from innovation

- Disseminates information through regular flow of bulletins, field (training) days, etc.

- Fosters small and family-owned orchards

Key problems

- APMB's cross-subsidisation of different varieties through a pool system masked quality differences amongst growers

- Difficulties in coping with infrastructure pressures brought about by newer, high-volume growers

- Wedded to homogeneous or general solutions, and assuming total uptake of changes

- Marketing inertia and slow response to changing conditions

- Difficulties in managing a greater diversity of growers – e.g. part-time and syndicated partnerships, corporations, and family companies

Source: Based on Grosvenor *et al.* 1995, Le Heron and Roche 1999 and McKenna *et al.* 2001.

The questions often posed within the industry are whether the producer-marketing board structure: has shifted or assisted innovation; has been a more or less effective marketing framework; and has provided higher or lower returns to the growers and/or industry as a whole. Table 5.1 summarises in greater detail the sentiments most often expressed about the APMB's worth as an institutional governance framework. Importantly, in New Zealand's two largest production regions – Hawke's Bay and Nelson – the APMB's popularity has remained high (McKenna 1999). In Hawke's Bay, the rapid and sustained increase in volume of fruit opened the possibility of increasing returns to scale, a situation recognised as conducive to industry innovation. This foundation for growth was stimulated by a land-use intensification process in the early 1980s (driven by assumed higher returns per hectare being capitalised into land prices) and reinforced by large scale planting of new varieties owing to investor perceptions of increasing returns for these and the absence of institutional barriers to entry.

In the late 1990s, amidst a more general furore over the possible abolition of the producer-marketing boards in New Zealand, the APMB had an opportunity to implement a shift in grower culture towards that of a low chemical production regime. The initiative to introduce Integrated Pest Control (IPC) reveals both the advantages and disadvantages inherent in the APMB framework (McKenna *et al.* 1998). Following earlier, hard-line decrees on packhouse hygiene regulations (required by the U.K. supermarkets) in 1991 and the Growsafe protocol aimed at sensitising growers to monitoring pesticide application levels in 1994-95 (Le Heron *et al.* 1997, Perry *et al.* 1997), the IPC project met with substantial resistance. While some of this was regionally specific – e.g. Hawke's Bay growers had to contend with more pest blooms than in Nelson – and entangled with anti- and pro-APMB sentiments, the fundamental difficulty remained that of managing an industry-wide shift in grower practices. In this case, it was to catch up with well-established conventions in competitor regions overseas, such as Northern Italy. The momentum of IPC was lost when ENZA faced severe export market conditions in the 1999 season and a resurgence of deregulation calls from privatisers in the industry. An internal restructuring of the APMB anticipated the passage of the Apple and Pear Industry Restructuring Act in 2000, and resulted in the formal separation of functions and the incorporation of the APMB into ENZA Ltd. Among its activities, this private company retains the APMB's production and post-harvest research programmes, which have otherwise been downgraded in New Zealand's public funding framework.

The experience of the apple industry under the producer-marketing board regime is illustrative of many aspects of the embeddedness model. Previous studies of the sector have identified the untraded interdependencies within the regional industry and cited these as essential factors in the export marketing success of the single international enterprise (Hayward *et al.* 1998, Le Heron and Roche 1996). In assessing the APMB's role in institutional development, three things stand out. First, the APMB enjoyed for many decades a remarkable level of grower confidence and trust. To most regional growers, the APMB structure was a safe haven. Whatever the industry income in any season, it was always distributed to growers – notwithstanding grumbles about the basis of distribution. Second, while relatively slow in implementing change (owing in part to the cooperative political framework that underpinned the APMB), the APMB was clearly responsible for developing, introducing and managing the market testing and growth of new varieties, and devised schemes to reward participants. However, despite the on-going advantages assured by the producer-marketing board system, the long-term ability of the industry to manage and foster innovation was unknown. For instance, it is possible that the local regimes that emerged in the two main regions reduced regional economies of scale and scope (McKenna *et al.* 2001). It is not known, however, whether the non-APMB alternatives envisaged by the deregulators would have yielded a greater level of commercialised innovation. Some clues lie in the proliferation of marketing ventures by some growers following the deregulation of the domestic market in 1997, and the subsequent further deregulation through the granting of export licenses by a new regulatory authority, the Apple and Pear Export Permits Committee. These developments sent shock waves through the regional growing community and gravely undermined the co-operative culture in Hawke's Bay. In the late 1990s, the once-sheltered New Zealand industry was becoming further exposed to international forces and integrated within an emerging global fresh produce complex (Hayward and Le Heron 2001). Furthermore, a salutary reminder of the vulnerability of New Zealand production to incursions by global capitalism occurred in July 2000 when growers suddenly learned that the investment companies Guinness Peat Group and FR Partners had taken a 40% stake in ENZA Ltd. (Stevenson 2000a). In effect, this opened the New Zealand (and Hawke's Bay) apple industry to corporate capital and full commercialisation. Further evidence of the cultural shift occurred with the replacement of ENZA's partially grower-appointed board of directors with one exclusively representing corporate investors and corporate growers (Stevenson 2000b).

Vegetable Industry: Corporate Networking, Intellectual Capital and Territoriality

The third example of institutional dynamics is vegetable processing. The Hawke's Bay region has been home to at least one major food processing company ever since J. Wattie Canneries ('*Wattie*', established in 1939) expanded rapidly to meet wartime demand[2]. At the beginning of the 1990s, Hawke's Bay's Wattie was part of the Australasian food conglomerate, Goodman Fielder Wattie (Le Heron 1990), and was steering a relatively independent course based on its traditional core area of fresh fruit and vegetable processing. While Wattie has operated continuously since establishment, attempts to operate a second sizeable food processing plant in the area have only been medium-term successes. The multinational, Unilever, operated from the 1940s until 1975; locally-owned Growers operated from 1978 until 1996; and the transnational McCain Foods started operations in 1996. Wattie has depended on supply from a loose and changing coalition of growers who were perennially dissatisfied with the company's contract prices and growing requirements (Hayward *et al.* 1998). The emerging focus was food exports, especially to Australia and to Asia. The 1992 sale of the Wattie division to the transnational food company H.J. Heinz, inserted the region's leading producer into a different corporate world. As part of this global food giant, Hawke's Bay was designated first as one of Heinz's eight core production areas (1996) and then two years later was redesignated as one of six production nodes for the company (Evening Post 1999). Heinz expects the region be the 'Garden of Asia' (McKenna *et al.* 1999) and this underpinned subsequent Heinz-Wattie investment in the region in processing facilities, research and development and human capital. Furthermore, in pursuit of Wattie's early 1990s commitment to organic production, Heinz-Wattie has continued to develop an organic food portfolio (Coombes and Campbell 1998).

The central issue of the 1990s has been the survival of Wattie as a processor in the face of Australian competition within the Heinz Corporation; with Dandenong, Victoria being an alternative production site (Pritchard 1995). In this process, the embeddedness of Wattie has been apparent as local government has cooperated with local management to portray the region as a low cost, innovative processor, with appropriate supply connections, and backed by a supportive local government sector. Hayward *et al.* (1996) show how Wattie's management enlisted the help of Hastings and Napier cities and the Port of Napier to support their internal bid within the Heinz corporation. The closure of Weddell's Tomoana meat works also released a substantial, fully serviced industrial site at a time when Heinz was evaluating Hawke's Bay as a site of further investment (Hayward *et al.* 1996). The Tomoana plant has been converted by Heinz-Wattie to the production of soups and sauces (The Press 1999). However, the site is also used for product innovation, partly owing to its small scale of operation. This specialisation as a development site and test bed is a particular niche within the Heinz international division of labour, and the role has brought new demands on local relationships. First, an expansion to product sourcing has been actively pursued. This has included the sourcing of products from adjacent regions, such as potatoes from Manawatu, peas from Gisborne and Manawatu, carrots from Ohakune, as well as locally-sourced beans. Second, professional and technical staff are being imported from outside the region. For instance, experts from Japan spend time on secondment in the region to assist in the production of traditional Japanese recipes using New Zealand ingredients. Third, the pursuit of the organic philosophy and encouragement of organic farming has occurred. However, Heinz-Wattie continues to experience difficulty in managing grower-processor relationships, which indicates the limited embeddedness that is inevitable for this local branch of a transnational corporation (Hayward *et al.* 1998). Nonetheless, the perception of New Zealand as clean and green, and potentially organic, is a viewpoint publicly stated in Heinz's Annual Reports and public statements in New Zealand, and is a more general part of the re-imaging of New Zealand horticulture (Le Heron and Roche 1999). Although difficult to confirm empirically, Heinz's assessment of the Hawke's Bay region as a site of production is likely to have rested at least in part on this perception.

Possibilities and Constraints in the Regional Structural Context

The chapter began by questioning whether food commodity chain interdependencies existed in Hawke's Bay. The review of the three major food production systems in the region has detailed relatively autonomous industry trajectories. The chains each have very different regional institutional structures. The lack of cross-chain interdependencies at the production, distribution and marketing levels meant restructuring in each of the co-existing chains was primarily played out with respect to each industry. Local government responses to the regionally cumulative impacts in each of the three industries have made little progress in vitalising the overall economic base, as the region amounts to three specialised and geographically proximate industries rather than an industrial complex constituted by local, growth-inducing interdependencies (Hayward and Le

Heron 1997). Indeed, Hayward *et al.* (1996) noted that neither a gap filling nor a networking strategy for the region had been extensively explored by regional institutions.

The three case studies throw light on two issues in the literature: the conditions under which networking might intensify and thereby assist growth processes and whether the global place of a regional industry in global commodity chains is affected by embedding or disembedding processes. In contrast to manufacturing sectors (such as in Gereffi 1999) where organisational learning is hypothesised to improve the positions of firms in international trade networks, the Hawke's Bay agricultural industries show no real congruence with a 'Gereffian' world. Instead, in each case, the regional industries have had a long history of participation in the global economy and display differentiated responses and fortunes.

The meat industry is an example of disengagement from the region. Despite the number of plants found in the region, local firms were unable to transcend supply competition and independent market channel development. For example, stock can be transported from the region's farms to processors elsewhere in the North Island (and often are when stock is in short supply and scarcity increases procurement competition). Any cross-over of the meat industry with horticulture appears to have been minimal; neither the Meat Producer Board nor national research institutions located elsewhere formed relationships that were in any sense advantageous during the rounds of industrial rationalisation that swept the industry from the mid-1980s onwards. From a regional standpoint, the Weddell collapse was sudden, total and irreversible. The later rationalisation and redirection of Richmond – first into a food company and then absorption within the Brierley Group – only further problematised the industry's enterprise composition in Hawke's Bay; although Richmond's strategies have opened up new market prospects for the region in high value food services and supply (New Zealand Herald 2001).

The apple complex, in contrast, has been far more encouraging in terms of institutional development. Coordination and control under the APMB regime led to a period of stable and reliable rules and conventions for the apple industry. Participation in the industry meant an acceptance by growers of a well-developed institutional environment. Until the 1980s, this was largely unquestioned. However, the appearance of new, large corporate growers and continued expansion of family orchards, coupled with the zero-entry costs for use of the APMB infrastructure, created a situation where the existing growers were in effect cross-subsidising the infrastructure investments needed to accommodate the newcomers. This internal dynamic coincided with wider calls in New Zealand for producer marketing board deregulation. Advocates of deregulation within the industry foresaw opportunities to gain low cost access to portions of ENZA's international markets. Nonetheless, the APMB (even in its reformed guise) still exercised an extraordinary influence on grower behaviour and retained the allegiance of 80-90% of growers. Not surprisingly, the APMB-privateer relationship has been a distraction for the industry but what is less clear is the extent to which innovation in the industry has changed, and the formerly cooperative and interdependent industry culture has been supplanted by a competitive, corporate structure. A number of the large corporate growers have maintained their support for the Board. Indeed, Hawke's Bay's own Eastern Equities, for instance, has retained two directorships in a recent boardroom coup (Stevenson 2000b). Their loyalty to the APMB may be partly explained by their specialisation in volume apple production of standard varieties, and their emphasis on field and packhouse efficiencies, which have accorded with ENZA's marketing strategies (Fox 2000). In contrast, the maverick and vocal deregulators contend that their fruit should be differentiated in the market place. The recent decision by the UK supermarket group, Tesco, to source a third of its New Zealand apple imports from an alternative supplier (300,000 cartons) is an indication of the global pressures being introduced through deregulation, and confirmation that New Zealand growers are still firmly entangled in power relations in which they are becoming increasingly marginalised (Dominion 2000). A further price collapse in 2000, on top of a marketing bungle in 1999, has strained the APMB's and ENZA's credibility as a viable regime for the industry. The majority of growers still subscribe to the APMB's political framework and organisational ethos, and the degree to which it can internalise procedures to handle fruit differentiated by ownership will probably determine its survival as an institution. However, while attention is directed primarily towards current earnings, the longer-term issues remain those of the next generation of innovations in varieties, orchard management, labour utilisation, packaging, imaging, cool chain efficiencies, and distribution economies. The increasing volume of output from the region will necessarily lead to a growth effect, even if a reconstructed and revitalised regional institutional structure is not created. The internal differentiation in the apple industry has problematised the placement of the region in the world apple complex. The crucial question for the region is whether the emergent groups of growers can retain their historically premier position.

The vegetable industry comes closest to the Gereffi upgrading model. The selection of the Hawke's Bay site by Heinz, as part of its global production strategy, is an indication that intra-corporate networking in a post-merger situation can have positive regional outcomes. The Heinz-Wattie complex has neither downsized nor closed. Instead, it has actually grown in the region through being realigned within an evolving corporate framework; leading to relocations of processing from Auckland and from Japan, and the acquisition of the Tomoana plant (Evening Post 2000). Solutions to supply variability have been developed: multi-area sourcing to reduce biophysical risk, and supplementary sourcing from countries such as Thailand and China (for tomato paste). Again, a long-term regional development issue continues to be the absence of food research facilities in Hawke's Bay and the limited nature of many educational and training options at the local polytechnic. Efforts to build alliances with tertiary providers have never advanced very far. Whereas in an earlier guise, the Wattie company contributed to the national, post-World War II development project in New Zealand, the present Heinz-Wattie complex is explicitly oriented to co-development of global markets within the company as part of a globalised, corporate logic. Thus, seemingly an instance of upgrading, the corporate network of Heinz appears to have released pre-existing potential within the Wattie enterprise through its insertion into a bigger and geographically denser corporate web.

Conclusions

The agri-processing economy of Hawke's Bay features a small number of sizeable processing enterprises in broadly interdependent relationships with a large set of farmers and growers. In unpacking the nature and dynamics of these interdependencies we have shown that tendencies to develop or undermine untraded interdependencies are complex. Our account has been highly selective, a simplification, and unquestionably sketchy. However, despite shortcomings, we believe our analysis supports the view that embedding and disembedding must be analysed as processes in structural contexts. We pluralise 'context' because most regions consist of a range of locally dominant components of commodity chains, and are set within national and international governance regimes.

Our first – and by no means exceptional – conclusion is that the co-ordination capacities of local actors are 'thickly' constrained by wider corporate and institutional networks at a variety of spatial scales. These constraints may be resisted locally or regarded as justification for alternative institutional solutions. Regardless of how they are perceived, the constraints pose resourcing problems for initiatives that are at variance with the expectations of key organisations. Constraints, in so far as they represent relationships in the circuits of the other enterprises, can also form opportunities; and through local efforts the opportunities might be recomposed into new relationships attuned to wider changes in the larger globalising economy. These dimensions may be very critical in how regions are inserted or reinserted into the global economy.

A second conclusion is that local actors, operating in the context of particular organisational and industry logics, are continually reconstructing business relationships. This necessarily involves the assessment of existing and potential relationships, and is a process that is thoroughly spatial in how it unfolds and also in terms of outcomes. Different gatekeeper organisations bring scope and content differences to the attempts at vertical or horizontal organisation. Organisational proximity – or more correctly, organisational accessibility – such as through merger, is perhaps a necessary condition in placing regional production within the global system.

A third conclusion, for Hawke's Bay at least, is that it is inappropriate to suggest that the region is an agri-processing district approximating to what is conventionally understood to be an 'industrial district'. The accounts of the three chains reveals the way political-economic actors in the region and elsewhere have constructed economic and regulatory spaces, generating layers of complexity that shape the spatial differentiation associated with the region.

This chapter has offered a relatively strong corrective to the notion that upgrading necessarily follows from engagement with global commodity chains. The focus on three different agricultural systems was a severe test of both the embedding and upgrading theses. Foremost, the regional industries exhibit different mechanisms by which learning has occurred, but the outcomes were often less than satisfactory from a regional perspective, and only partly support the upgrading thesis. A more appropriate conceptualisation is that of reinsertion, while a pre-condition for possible upgrading, does not imply sequential processes. In a similar vein, the variability in outcomes suggests that it would be hard to predict typical trajectories in terms of export roles. For the industries studied at least, there is poor support for a model sequence of development.

Acknowledgements

David Hayward and Richard Le Heron would like to acknowledge the support of the Auckland University Research Committee, project ID 3420177 ('Hawkes Bay Infrastructure'). Richard Le Heron gratefully acknowledges the support of the Foundation of Research Science and Technology project UoAX0009. Christina Stringer gratefully acknowledges the support of a University of Auckland doctoral scholarship.

Notes

1　Details of the company behaviour, the North Island focus of company stock purchasing, anti-union tactics, inter-company rivalries and the non-involvement of communities is covered in Le Heron (1991a, 1992).

2　See Le Heron and Warr (1976) for an account of the Wattie and Unilever strategies in the region until the mid-1970s and Cooper *et al.* (1995) for discussion of the mid-1990s.

References

Blunden, G. (1996), 'Corporatisation and the producer marketing boards', in R. Le Heron and E. Pawson (eds), *Changing Places – New Zealand in the Nineties*, Longman Paul, Auckland, pp. 126-129.

Capello, R. (1999), 'Spatial transfer of knowledge which technology milieux: Learning versus collective learning processes', *Regional Studies*, vol. 33, no. 4, pp. 353-365.

Chistopherson, S. (1999), 'Rules as resources: How market governance regimes influence firm networks', in T. Barnes and M. Gertler (eds), *The New Industrial Geography: Regions, Regulation and Institutions*, Routledge, London, pp. 155-175.

Coombes, B. and Campbell, H. (1998), 'Dependent reproduction of alternative modes of agriculture: Organic farming in New Zealand', *Sociologia Ruralis*, vol. 38, pp. 127-145.

Cooper, I., Perry, M., Le Heron, R.B. and Hayward, D.J. (1995), *Business Networking and Exporting: A Profile of Networking and Networks in the Apple, Asparagus, Grape, Squash and Tomato Sectors of the Hawke's Bay Horticultural Complex*, Occasional Paper 31, Department of Geography, University of Auckland.

Curtis, B. (1999), 'Constructing markets: Governance in the meat and dairy industries of New Zealand', *Rural Society*, vol. 9, no. 2, pp. 491-503.

Dominion (2000), 'Apple grower cuts deal without ENZA', Wellington, 12 February, p. 9.

Evening Post (1999), '"Garden" NZ key to Heinz Plan', Wellington, 19 February, p. 11.

Evening Post (2000), 'Heinz quits Japan', Wellington, 10 February, p. 12.

Fox, A. (2000), 'Enza at core of export war', *The Dominion*, Wellington, 16 February, p. 23.

Gereffi, G. (1999), 'International trade and industrial upgrading in the apparel commodity chain', *Journal of International Economics*, vol. 48, pp. 37-70.

Grosvenor, S., Le Heron, R.B. and Roche, M. (1995), 'Sustainability, corporate growers, regionalisation and Pacific Asia links in the Tasmanian and Hawke's Bay apple industries', *Australian Geographer*, vol. 26, no. 2, pp. 163-172.

Hayward, D.J. and Le Heron, R.B. (1997), 'Infrastructure issues and regional competitiveness in Hawke's Bay, New Zealand', Paper presented at the Institute of Australian Geographers and New Zealand Geographical Society 2nd Joint Conference, University of Tasmania, January.

92 *Embedded Enterprise and Social Capital*

Hayward, D.J. and Le Heron, R.B. (2001), 'Horticultural reform in the European Union and New Zealand: Further developments towards a global fresh fruit and vegetable complex', unpublished paper.

Hayward, D.J., Le Heron, R.B., Perry, M. and Cooper, I. (1998), 'Networking, technology and governance: Lessons from New Zealand horticulture', *Environment and Planning A*, vol. 30, pp. 2025-2040.

Hayward, D.J., Le Heron, R.B., Stringer, C. and Le Heron, K.L. (1996), *Hawke's Bay Industrial Development Project*, Department of Geography, University of Auckland.

Hwang, J.S. and Choo, S. (2000), *Characteristics and Development of Industrial Districts – The Case of Software Clusters in Seoul, South Korea*, Paper presented at the International Geographical Union commission on The Organisation of Industrial Space conference on 'Knowledge, Industry and Environment', Dongguan, China, August.

Le Heron, R.B. (1980), 'The diversified corporation and development strategy – New Zealand's experience', *Regional Studies*, vol. 14, no. 3, pp. 201-218.

Le Heron, R.B. (1989a), 'A political economy perspective on the expansion of New Zealand livestock farming, 1960-1984. Part 1: Agricultural policy', *Journal of Rural Studies*, vol. 5, no. 1, pp. 17-32.

Le Heron, R.B. (1989b), 'A political economy perspective on the expansion of New Zealand livestock farming, 1960-1984. Part II: Farmer responses – aggregate evidence and implications', *Journal of Rural Studies*, vol. 5, no. 1, pp. 33-41.

Le Heron, R.B. (1990), 'Good food worldwide? Internationalisation and performance of Goodman Fielder Wattie Ltd', in M. de Smidt and E. Wever (eds), *The Corporate Firm in a Changing World Economy*, Routledge, London, pp. 100-119.

Le Heron, R.B. (1991a), 'New Zealand's export meat freezing industry: political dilemmas and spatial impacts', in D. Rich and G.J.R Linge (eds), *The State and the Spatial Management of Industrial Change*, Routledge, London, pp. 108-127.

Le Heron, R.B. (1991b), 'Corporate raiders and industrial restructuring: The case of Brierley Investments Ltd', in E.W. Schamp, G.J.R. Linge and C.M. Rogerson (eds), *Finance, Industry and Institutional Change*, Walter de Gruyter, Berlin, pp. 83-102.

Le Heron, R.B. (1992), 'Meat freezing industry restructuring', in S. Britton, R.B. Le Heron and E. Pawson (eds), *Changing Places in New Zealand. A Geography of Restructuring*, New Zealand Geographical Society, Christchurch, pp. 107-113.

Le Heron, R.B. and Pawson, E. (eds) (1996), *Changing Places: New Zealand in the Nineties*, Longman Paul, Auckland.

Le Heron, R.B. and Roche, M. (1996), 'Globalisation, sustainability and apple orcharding, Hawke's Bay, New Zealand', *Economic Geography*, vol. 72, no. 4, pp. 416-432.

Le Heron, R.B. and Roche, M. (1999), 'Rapid reregulation, agricultural restructuring and the reimaging of agriculture in New Zealand', *Rural Sociology*, vol. 64, no. 2, pp. 203-218.

Le Heron, R.B. and Warr, E.C.R. (1976), 'Corporate organisation, corporate strategy and agribusiness development in New Zealand: An introductory study with particular reference to the fruit and vegetable processing industry', *New Zealand Geographer*, vol. 32, pp. 1-16.

Le Heron, R.B., Cooper, I., Hayward, D.J. and Perry, M. (1997), 'Commodity system governance by quality management: A New Zealand discourse', in M. Taylor and S. Conti (eds), *Interdependent and Uneven Development*, Ashgate, Aldershot, pp. 81-100.

Lynch, B. (1996), 'Meat industry restructuring', in R.B. Le Heron and E. Pawson (eds), *Changing Places, New Zealand in the Nineties*, Longman Paul, Auckland, pp. 142-147.

McKenna, M. (1999), 'The emperor has no clothes – geographies of deregulation', *New Zealand Geographer*, vol. 55, no. 1, pp. 59-62.

McKenna, M., Roche, M. and Le Heron, R.B. (1998), 'Sustaining the fruits of labour: A comparative localities analysis of the integrated fruit production programme in New Zealand's apple industry', *Journal of Rural Studies*, vol. 14, no. 4, pp. 393-409.

McKenna, M., Roche, M. and Le Heron, R.B. (1999), 'H.J. Heinz and global gardens: Creating quality, leveraging localities', *International Journal of Sociology of Agriculture and Food*, vol. 8, pp. 35-51.

McKenna, M., Le Heron, R.B. and Roche, M. (2001), 'Living local, growing global: Renegotiating localities in New Zealand's pipfruit sector', *Geoforum*, in press.

Markusen, A. (1999), 'Sticky places in slippery space: A typology of industrial districts', in T. Barnes and M. Gertler (eds), *The New Industrial Geography: Regions, Regulation and Institutions*, Routledge, London, pp. 98-124.

New Zealand Herald (2001), 'Richmond company advertisement: "Food simply the best...Gourmet Direct"', Auckland, 22 January, p. D2.

Peterson, J. and Holland, P. (1996), 'Marketing New Zealand's apple industry', in R.B. Le Heron and E. Pawson (eds), *Changing Places: New Zealand in the Nineties*, Longman Paul, Auckland, pp. 132-136.

Perry, M., Davidson, C. and Hill, R. (1995), *Reform at Work*, Longman Paul, Auckland.

Perry, M., Le Heron, R.B., Hayward, D.J. and Cooper, I. (1997), 'Growing disciplines through total quality management in a New Zealand horticultural industry' *Journal of Rural Studies*, vol. 13, no. 3, pp. 289-304.

The Press (1999), 'Heinz Wattie expands', Christchurch, 20 August, p. 17.

Pritchard, W. (1995), *Uneven Globalization: The Restructuring of the Australian Dairy and Vegetable Processing Sectors*, unpublished PhD thesis, The University of Sydney.

Stevenson, P. (2000a), 'Corporates wage battle for slice of apple pie', *New Zealand Herald*, Auckland, 22 July, p. E1.

Stevenson, P. (2000b), 'New broom sweeps away Enza board', *New Zealand Herald*, Auckland, 29 August, p. C3.

Storper, M. (1992), 'The limits to globalisation: Technology districts and international trade', *Economic Geography*, vol. 68, no. 1, pp. 60-93.

Stringer, C. (1999), *New Zealand's Relations with Northeast Asia: Links and Interactions Under Globalisation*, unpublished PhD thesis, University of Auckland.

Taylor, M.J. (1998), 'Enterprise, power and embeddedness: An empirical exploration', Paper presented at the International Geographical Union commission on The Organisation of Industrial Space conference on 'SMEs and Regional Development', Seville, August.

Chapter 6

Weakening Ties: The Embeddedness of Small UK Electronics Firms

Sara Openshaw and Michael Taylor

Introduction

The economic embeddedness model applied to industrialised countries
envisages growth in jobs, enterprise and output being fuelled by cooperation between co-located enterprises, built on inter-firm trust,
reciprocity and loyalty. It is a very persuasive model that has been
elaborated in detail in an extensive literature (Braczyk *et al.* 1998, Cooke
and Morgan 1998, Storper 1997, Taylor and Conti 1997, Maskell *et al.*
1998, Maskell and Malmberg 1999). Here, it is suggested, is an essential
element of the 'knowledge economy', built on social relationships, repeated
business, and the mobilisation of tacit and codified knowledge that
combine to create a dynamic learning process (Asheim and Isaksen 2000,
Lundvall and Johnson 1994). Within such a milieu, these social
relationships embed principally small firms into networks of weak and
easily reconfigured buyer-supplier transactions, creating 'social capital'
(Bordieu 1980, Coleman 1988, Putnam 1993, Portes and Sensenbrenner
1993, Fukuyama 1995, Woolcock 1998, Fedderke *et al.* 1999). And, like
money capital, social capital is reckoned to be vital for the creation,
sustenance and development of local economic activity. In this
'atmosphere' of goodwill and social cohesion, the virtuous relationships of
Porter's (1990, 1998) development diamond are said to enhance local
invention and innovation, to create jobs, to promote entrepreneurship and to
raise local productivity to globally competitive levels, when the
circumstances are right.

Through repetition, however, this model has become a mantra, and
'proof' has come from assertion. Repetition has replaced replication. The

most common methodology used to 'test' the model has been to select 'successful' regions and places and then to explore the nature of inter-firm relationships and institutional structures within them. The mechanisms that are found are assumed to be the mechanisms of success (Staber 1996). Such an assumption ignores the possibility that current 'success' might be a consequence of now extinguished processes, and that current processes might be relict, irrelevant or even destructive. Quite simply, the processes postulated in the embeddedness model of local economic growth are timeless and, being timeless, they obscure and confound any appreciation of the directions of causality in processes and mechanisms of local economic development and change.

The purpose of the analysis in this chapter is to begin empirically to unpack the direction and nature of causality between embeddedness and local growth. Here, local growth is treated narrowly as new firm formation. The empirical foundation of the study is a set of 11 case studies of small electronics firms in Poole, in the UK[1]. This is a geographic sub-unit of a much larger defence-related concentration of functionally integrated electronics firms in a region stretching from Dorest across southern Hampshire, including the urban centres of Romsey, Southampton, Eastleigh, Fareham, Portsmouth and Havant. The analysis attempts to associate the current nature of firms' buyer-supplier relationships with the manner in which these enterprises were first created and have subsequently evolved. What is evident from the analyses is that the circumstances that led to the creation of this sample of small firms are quite different from the circumstances under which they must now operate. The study is exploratory. It does not attempt to measure the extent to which different processes operate within the electronics industry at the present time. Its aim is simply to shed light on the nature of the processes that operate within and between firms in this sector. As such, it is a qualitative exploration of the nature of dynamic processes of local economic change.

The discussion of this chapter is developed in three stages. First, the origins of the surveyed electronics firms are examined to explore the characteristics of the economic environment within which they operate, the time-specific nature of the social and economic processes that operate within it, and dynamic of enterprise embeddedness. Second, the nature of the links firms have with their customers is explored to reflect on the embeddedness of these forward ties in the electronics supply chain. Third, the nature and extent of firms' supplier linkages is reviewed in a similar manner to look back down the supply chain. The merging of these three stands of analysis sheds light, in a preliminary, way on the dynamics of enterprise embeddedness in this South Coast cluster of electronics firms.

The number of firms interviewed is however small, and the conclusions that can be drawn can be no more than preliminary.

Embeddedness and New Firm Formation

The interviews with electronics firms in Dorset suggest very strongly that new firms are, for the most part, spin-offs from major defence contractors, as skilled employees branch out on their own to become sub-contractors within a defence-related supply chain. That supply chain ties together a cluster of defence related manufacturers in the coastal cities of Dorset and Hampshire that has a history going back centuries, but in terms of electronics goes back to the Second World War. As it was put by one firm:

> The electronics [industry] grew out of the defence industry. ... So, basically all these companies round here have been fed off ... second [and] third generation companies. I think that's why they're here. And, its navy orientated, with local Portsmouth way. (DM4)

Subsequently, these small electronics ventures have been bought, sold, expanded and closed as the people that run them have prospered, failed, lost interest and aged.

Spin-off is the basis of most new firm formation. Indeed, six of the surveyed businesses in this study have involved spin-off in their foundation: spin-offs from defence contractors, from electronics sub-contractors and, in one case, from a university. Firm PA1 is an example of the process (see Figure 6.1). The founder of firm PA1 originally worked for Plessey, a large electronics firm in Poole (now part of Seimens). When a group of engineers broke away from Plessey to form a design and manufacturing firm, they asked the founder to assemble printed circuit boards (PCBs) for extra cash in his spare time. He agreed, and when he realised there were substantial profits to be made, he set up his own business. Close, local buyer-supplier relationships have been at the heart of this firm's success, and these are shown in Figure 6.1 in the ties with DM'A' and the portfolio of firms centred on firm DM7. Since the firm was set up in 1972, his sister, her husband and his daughter have bought into the company, and were due to assume control of the business on the founder's retirement in 2000. Succession, however, does not always stay in the family and firms PA2 and PM2 have been bought from their founders by unrelated interests who still run them.

Coupled with spin-off in the creation of new firms has been the identification of gaps in the market. In the case of firm PA1, the market came to the founder. The founder of DM5 saw an opening in the success by working closely with a printer manufacturers. Another firm (firm DM6), was started when its founder was working for a producer of small electronic control systems. That company did not want to move into market related to printer technology and bar-coding, and secured the new firm's production of larger controls so this individual used his experience and

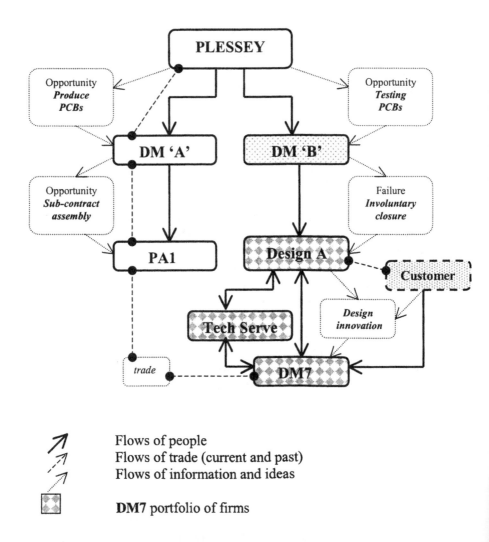

Figure 6.1 Spin-offs and embeddedness (for explanation see text)

expertise and the opportunity of a low rent subsidised by the Rural Development Commission to open his own business.

Some small electronics firms are not such simple, single site operations run by sole proprietors, partnerships and family partnerships. A significant number of the surveyed firms have evolved to become parts of more complex structures of ownership controlled by entrepreneurs and coalitions of entrepreneurs (Taylor 1999, Rosa and Scott, 1999). In these circumstances, entrepreneurs have interests in two or more businesses, and five of the firms in the current study fall into this category. Thus, firm PM1 was bought from the receivers in 1994 by the owner of a Midlands firm which now operates at least two businesses. Similarly, when firm DM1 got into financial difficulties in the mid-1980s, it was sold on to one of its customers by a venture capital firm. The principal owners of the two firms are a husband and wife. Further complicating the picture, the Chairman of DM1 has invested in both DM1 and the company that owns it, but plays no management role in either firm. He is in effect an investment angel with, as far as the informant to the survey knew, additional business interests and links through directorships into corporate sector businesses and other activities in the City of London. In firm DM1, therefore, two sets of portfolio entrepreneurs intersect in their pursuit of personal wealth creation. In a similar vein, the owners of firms DM3, DM4 and DM7 have business interests beyond the firms surveyed in this study, and the history and origins of firm DM7 provide a different perspective on the complex processes involved in portfolio entrepreneurship (see Figure 6.1).

Firm DM7 is principally owned by a man and his wife, along with a number of private investors. They also own a company that offers technical services, and are involved in a computer aided design business. The husband had been a design engineer at Plessey who left to establish his own business to design and make a product to test printed circuit boards. Although initially successful, the firm was forced to close by its bank (Figure 6.1). Later, the entrepreneur opened another company offering design engineering services to implement manufacturable designs. One idea brought to this company concerned computer networking. The entrepreneur and the originator of the idea tested the viability of the idea and used it to found firm DM7 in 1993. The person who had the original idea has recently been bought out of firm DM7 as he wanted to concentrate on his own business interests. What this example demonstrates, as with firm DM1, is the way in which coalitions of people are created and dissolved to run and expand businesses for the purpose of personal wealth creation. For the investor, portfolio entrepreneurship stabilises wealth creation investor and hedges against failure. It is hardly surprising, therefore, that it is so common among the electronics firms surveyed for this study.

It is also clear from the interviews with electronics firms that the environment that has spawned new firms as sub-contractors to the South Coast defence industries is strongly time-specific. In the environment of the 1970s, at the height of the Cold War, large defence contractors worked on rolling defence programmes, principally for the Ministry of Defence. The programme contracts were drawn on a cost-plus basis, and provided great security and stability in the sector. The defence contractors retained large workforces and trained large numbers of technical experts. In this cosy environment, there were plenty of opportunities for skilled employees to spin-off businesses that could sub-contract work from the principal defence contractors (i.e. firms PA1 and the forerunner of DM7).

But, in the 1980s and subsequent years, the environment became more hostile and competitive. Change came on at least three fronts. First, there were defence cuts that came in the early 1980s following the election of a Conservative government. Through 1981 and 1982, cost-plus contracting was scrapped. Workers were laid off by principal contractors (firm DM4), and sub-contract electronics assembly work expanded. Now small firm start-ups were as much a refuge from unemployment as vehicles to exploit market opportunities. Second, to cut costs, defence contractors began to down-size and out-source from the beginning of the 1980s. To a certain extent, this restructuring was also part of a wider international management fashion to cope with globalising economic pressures. However, in the UK, defence procurement arrangements saw a shift to project-based contracting rather than rolling programmes. There was, as a result, a 'shuffling of the pack at the top' (firm DM4) as the large defence contractors restructured to cope with change. The assets of corporations like Plessey, Siemens, Marconi, GEC, Racal, British Aerospace, Vosper Thornycroft and others were rearranged through takeover, merger, rationalisation and re-engineering. Third, competition in the electronics industry has intensified, especially since the mid-1990s, forcing further change. 'This has been coming for the last five years. The Far East has hit us from every side, including competition' (firm DM3). Margins have been pared to the bone and firms have begun to explore markets outside the UK, in Europe (firm DM7).

A reconfigured supply chain has emerged in which the defence contractors work on projects with no guarantee of future work. They put out work only to the largest sub-contractors (firm DM4), pushing the small firms to a second tier and forcing them to look beyond defence work to electronics work in domestic fields like telecommunications, domestic appliances and so on. Workflow is now increasingly 'spiked', so that 'the numbers are OK, but the security's gone ... It doesn't feel real. You lack the confidence' (firm PM1). As workflow has become less predictable, the first tier sub-contractors are retaining or competing for the work once put out to second tier sub-contractors:

> Big circuit board companies are going to get short of work, or they already are and they are going to move down the ladder and start taking the medium-sized companies' work, who will then have to look further down. (firm PM2)

At the bottom of the sub-contract system the labour market has been destabilised. At this level, the PCB assemblers, the 'board stuffers', now offer increasing impermanent, insecure jobs:

> A lot of board stuffers employ staff, and the staff go up and down, ... so PCB staff get fluxed about. There are agencies that employ or provide short-term work for these people, and they move around. Everyone has to have flexibility in this business, you know, for competitive reasons or the nature of the work being project orientated. (firm DM4)

The inference is, therefore, that the economic environment that created small electronics firms in the decades from the 1950s to the 1970s has gone – and that was the environment that created seven of the eleven firms surveyed in this study. The more hostile environment that has developed through the 1980s and 1990s has created a new dynamic among electronics SMEs. Their existence has become more precarious. They are more frequently a refuge from unemployment. And, of necessity, they are turning away from the defence industries that have for so long offered them sustenance. In short, the processes of enterprise embeddedness that have fostered the growth of electronics SMEs in the coastal towns of Dorset and South Hampshire are in a state of radical change. That change is currently evident in the buyer and supplier relationships of the firms in the industry. The nature of those relationships is the subject of the next section of the chapter.

Inter-Firm Relationships

The extent to which small electronics firms are currently embedded in the socially constructed networks of local economies can be inferred from the nature and form of their buyer supplier relationships. Loyalty, trust and reputation are at the heart of these social relationships, while a preoccupation with price, the problems of lock-in and the exercise of power tend to destroy them.

Relationships with Customers

Most small electronics firms surveyed as part of the present study do not have contracts with customers, but transact business on single orders. Those orders, however, may continue for a considerable number of years, building the trust and loyalty and reciprocal relationships that are said to be the foundation of local embeddedness and the creation of local social capital. Here, at first sight, is an empirical example of the strength of weak ties.

Though price is important in shaping customer relationships, quality is important too. In this context, a design and manufacturing firm made the point that as their quality improved, and improved substantially in recent years, they have 'managed to persuade customers that price isn't really the issue, what really matters is the quality' (firm DM6).

Loyalty is at the heart of the surveyed firms' customer relationships. For one (firm PA1), the founder's relationship with a former work colleague who also set up his own firm has produced a continuous stream of orders over 30 years. For another (firm PA2), most of its customers have remained the same for 15 years or more, to the extent that 'it's rare we pick up new customers … generally because we are too busy'. For a third firm (firm PM2), loyalty has meant bringing customers with him from London when he relocated to Poole in the mid-1980s. Loyalty also means repeat business, even though it may be based on single orders. Repeat business builds trust between firms, as suppliers become familiar with designs and the requirements of their customers. Trust goes as far as one surveyed firm (firm DM2) despatching goods to customers on approval before invoices are sent.

It is clear, therefore, that small electronics firms build strong reputations. To build their reputations, they prefer to deal with engineers rather than buyers 'because they appreciate what you do. [Buyers] … are only interested in price. [T]hey just say that they allow you so much money, so that's what you're going to get, take it or leave it' (firm PA1). Once built, those reputations help them through recessions (firm PA1). They also help them cut costs. 'Your sales budget is lower if your reputation's good, because you're selling by word-of-mouth' (firm DM5). Reputations are also recognised and referenced by potential customers (firm PM1). Nevertheless, most of the surveyed firms continue to use standard marketing techniques to attract new customers. These range from cold calling to tele-sales, mail shots and direct targeting through trade shows, trade directories, web sites, Yellow Pages and Thompson's Directory. To some extent this is 'flag waving' (firm DM4), but even the cold calling used by two surveyed firms requires a point of entry through a specific person.

But, embedded as these customer relationships might at first appear, they also have a more exploitative side. An important theme in the interviews is that the electronics firms' relationships with their customers are relationships of service where, faced with strong competition, quality of product and service are often more important than price. However, there is a very fine line between 'service' and 'subservience', and subservience is very apparent owing to the powerlessness of these small firms. As one respondent pointed out:

> Every customer has their own rule book, and you have to treat each one differently and say, well, that's the way they want it done and that's the way we have to do it. And we get that amount of aggravation from that customer, and it's different from the aggravation from that customer. But, we learn what the rules are, ... and it's very frustrating ... but at the end of the day we get £10,000 of work from them a year. (firm DM6)

Clearly, the customers have the power in a trading relationship and, as the majority of respondents pointed out, that power is positively correlated with the direction of the flow of money. 'The one who is parting with the money has the power over the one who is receiving the money' (firm DM3). And, the only way to avoid subservience and lock-in is to avoid any one customer monopolising a firm's output. So, one firm attempts to avoid the 'all your eggs in one basket syndrome' (firm DM1), while another says, 'we've spread ourselves out, so we're now supplying a customer who might only take 1 per cent of our turnover' (firm DM5).

Lock-in is clearly a threat to small electronics firms. Purchasing orders that span 3 to 5 years (firm DM1) offer a certain amount of stability, but contracts have problems. As one respondent remarked about a customer contract:

> They think that if they give us a 12 month contract, which they assess every 3 months, it keeps the price down for them, because it means we're going to turn over 'x' amount of products in that year. They guarantee to buy that many products in a year. So it gives us a bit of security but it means they get a cheap job. (firm PA2)

'Preferred supplier' status, enjoyed by four of the surveyed firms, holds the same problems of price and potential lock-in:

> We've got this particular customer that ... knocks us down as cheap as we can get ... and then gives us lots of work and says he's going to flood us out. (firm PA1)

Focusing on price though, does not breed loyalty:

> Then, the next we know, he's tried someone else. Then, he gets a problem like quality and he'll come back to us. We know what's going on – he doesn't tell us what's going on. (firm PA1)

What is very clear from the interviews with small electronics firms is that the market environment within which they operate has changed rapidly and substantially in recent years. At the core of this change is the manner in which the corporate sector drivers of this industry now choose to do business. What they are creating is a multi-tiered sub-contracting system in which cost cutting and managerialism are destroying the embedded relationships that have created small electronics ventures in the past (see the arguments developed in Christensen (2000) and Johannisson (2000)). The principal market for the output of the small printed circuit board makers and design and manufacturing firms covered in this study are the corporate defence contractors such as Plessey, British Aerospace and GEC/Marconi. The way they now do business is well explained by one assembler of printed circuit boards:

> Where we used to get contracts from Plessey direct, now that Plessey has been taken over by Siemens, which is a German company, they will only put contracts out to firms perhaps 5 or 6 times bigger than us. So, they put the work out to them and they, in turn, sub-contract to us. So, we've gone back a line. (firm PA1)

Then, those large first-tier sub-contactors:

> ... hire and fire at will ... [S]ometimes they come along and they keep us busy for about 6 months at a time, and then it all stops. And then we may not see them again for a further 6 months or even 9 months, depending on if they get huge contracts. Then, swamp the area with work, and ... when it's been finished it all goes quiet. (firm PA1)

These changes have had two effects. First, they are increasing the formality of small firms' customer relationships with large firms, and second, they are associated with corporate strategies to increase input flexibility and yet maintain quality. In recent years, there has been an increase in demand for national and internationally recognised quality standards. These range from BSI 9000, which monitors procedures within companies, to Investors in People (IIP), which ensures that employers are aware of employees' training needs and that they adhere to training schedules. What these mechanisms of managerialism are doing is imposing a layer of bureaucracy on previously informal, socially guided relationships, blighting the trust and reciprocity that had previously been vital to them.

Coupled with this drive for flexibility is the perception amongst small electronics sub-contractors that customers now expect products to be manufactured a lot quicker than in the past. Orders that had formerly taken 10 to 12 weeks to complete are now expected to produce the first batches of product in less than four weeks. It is the pressures of de-stocking and the drive for increased efficiency associated with flexible production that is shortening lead times. Now, as a result of these changes, there is a general feeling of instability among the small firms in the electronics industry, and a suspicion that the larger first-tier firms will take work from them.

A number of significant conclusions can be drawn from this analysis. First, there are some elements of embedded economic relationships in small electronics firms' transactions with customers, drawing on trust, loyalty, reputation and a concern for quality. Second, those embedded relationships have been undermined by the changing strategies on purchasing and sub-contracting being pursued by TNCs and large corporations especially in the defence industries. Now those corporations deal only with larger suppliers on a just-in-time basis to cut costs and enhance the speed and efficiency of their own operations. They are also formalising and bureaucratising transactions by demanding recognised quality standards. Indeed, the small electronics firms covered in this study are pushed into subordinate second-tier positions to serve an increasingly unstable demand. Third, the embeddedness of small firms' relationships is being further eroded by the exercise of power by first-tier sub-contractors. Their power is over prices, the placement of orders and the conditions they are able to impose on deliveries. Taken together, these three sets of processes suggest that the embedded system that existed until recently in the defence related cluster spanning Dorset and southern Hampshire, is now being eroded. The processes of erosion, it can be argued, are destroying the conditions that once created new small firms as spin-offs from large corporations.

Relationships with Suppliers

In contrast to the weakening embeddedness of transactions with their customers, the current relationships between the surveyed electronics firms and their suppliers are much more strongly socially constructed. Indeed, trust, loyalty, close working relationships, and recommendations and referrals within networks, underpin the majority of the working relationships between these firms and their suppliers.

Eight of the surveyed firms had no formal contracts with their suppliers. Even the remaining three (firms DM2, DM3 and DM4), who are BSI 9000 registered and therefore must use formal contracts to comply with regulations, sometimes buy 'on a looser description basis' (firm DM3). The general attitude over dealings with suppliers is 'if you can't work on a loose tie then there's not much point. If you get involved in legal battles you're in a no-win situation – both sides' (firm DM6). 'We could tie them [suppliers] in knots legally, ... but its not going to benefit the relationship, so what's the point of putting it all in writing?' (firm DM5). Indeed, the same respondent went on to say that '[A] contract would tie us up, I suppose', implying that contracts would limit their own flexibility.

Loyalty is a significant aspect of electronics firms' relationships with suppliers, but tempered with an appreciation of price and competition. Four respondents (firms DM2, DM5, DM7 and PM1) have between six and eight suppliers each, and use them all. They build close ties with each supplier but retain their power as buyers to swap between suppliers. Some firms, however, have no choice in what suppliers they use as there is only one (e.g. firm DM7), while others (firms DM3 and DM6) choose to use only one supplier for work they sub-contract out. These last two firms are looking for the loyalty that comes from exclusivity so they can 'exert a bit of pressure' (firm DM6) when they have to. They want closeness to ensure quality and service because 'price is almost a non-factor in the electronics industry' (firm DM3). Most of the interviewed respondents say they pay a premium for their inputs to ensure quality and service. This has created many longstanding relationships with suppliers in the electronics industry that reduce the chances of mistakes in production and allows both buyer and supplier to react quickly to market changes.

In this environment of trust and loyalty in input transactions, it is of no surprise that a number of firms work closely with their suppliers, and that their 'businesses are knitted together' (firm DM3). This closeness can assume a number of forms, and three were mentioned by respondents: technical collaboration, financing and the prioritisation of work. A supplier of chemicals of a manufacturer of printed circuit boards, for example, works on-site with this firm 'developing new processes and analysing processes' (firm PM1). A design and manufacturing firm (firm DM7) has gone so far as to provide an interest-free loan of £10,000 to a PCB assembler it buys from so that the assembler is able to buy new equipment to produce their PCBs more cheaply and efficiently. A number of firms collaborate with suppliers so that the items being made for them are given priority in production schedules – but not on a regular basis. The reason for this closeness was summarised by one firm. 'Everyone has a huge vested interest in providing service both ways, and restoring that level of service should be everyone's number one priority' (firm DM3). This closeness in commercial relationships can last for years, but it was also remarked that familiarity can create its own problems.

Close ties with suppliers also means the use of local suppliers, and a majority of the businesspeople interviewed stressed that they use local suppliers where possible. There are good reasons for this. Problems can be resolved more easily and technical back-up can be guaranteed when the supplier is local. These benefits more than off-set the cheaper sub-contracting and parts supply that can be obtained from foreign suppliers. Nevertheless, three design and manufacturing firms are beginning to use overseas suppliers in Europe (Italy) and the Far East (Hong Kong) (firms DM3, DM5 and DM7).

There is evidence from the interviews that electronics firms close relationships with their suppliers in this part of the UK constitute a 'third-party referral network' (Taylor, 1999). In such networks, people and companies that might be able to meet the needs of a particular business (the third parties) are recommended in conversations between people working in other businesses. In the present study, these arrangements are evident when the majority of respondents maintain that having a good reputation brings work through word-of-mouth. One firm (firm DM6), for example, decided to look for a new sub-contractor because its existing sub-contractor was persistently failing to meet deadlines. On mentioning this problem to their PCB supplier, that supplier said 'he knew a very good man who won't let you down – he's very good. He's not the cheapest, but he's good'. Other respondents remarked on the same mechanism and its reliability. 'Its amazing that ... in these industries people do use word-of-mouth, but its something you can always rely on, something you can always trust' (firm DM2). There are obvious benefits in this system that arise from putting issues, problems and solutions at a distance. Two parties discuss an issue but do not seek a solution between themselves. In this way the issue or opportunity is discussed at arm's-length. Equally, the solution through a third party means that the relationship between the first two parties is put in less jeopardy.

Nevertheless, three of the surveyed firms (firms PM1, PM2 and DM2) deal with their suppliers on price alone. Socially constructed supplier relationships are not part of the way they do business. In part, this is a situation created by standardisation in printed circuit board manufacture, and the attitudes of this sub-group are well summarised by a PCB manufacturer:

> If we've got two suppliers, one we've been dealing with for ten years and one for ten minutes, … if one were more expensive, we would swap right away. … [A]s far as our suppliers are concerned, they more or less supply the same product, albeit under different names, but it is basically the same …[and]… quality is not really an issue. (firm PM2)

What is clear from this analysis of small electronics firms' supplier relationships is that, except where inputs are standardised, those relationships are strongly locally embedded They involve trust and reciprocity in good measure and have built into third-party referral networks that maintain the flow and redistribution of work between firms. Those same relationships have been suggested as channelling technology and finance in certain circumstances. They have also been recognised as enabling production schedules to be 'adjusted', smoothing workflows elsewhere in the local supply chain. These relationships stand in stark contrast to those that the same firms enjoy with their customers. Nevertheless, as more parts supply and sub-contracting is moved offshore, as is happening in three of the surveyed firms, these close supplier relationships may begin to break down much as their customer relationships have.

Conclusion

A very clear picture of the dynamics of enterprise embeddedness emerges from this analysis of small electronics firms. The picture that emerges of new firm formation processes is one that is essentially time-specific and time-dependent. Government funded defence contracting in the decades following the Second World War, created a protected environment that fostered close inter-firm relationships in the electronics sector that was conducive to new firm formation. From the beginning of the 1980s, this environment changed. A leaner approach to defence procurement, international competition and corporate sector restructuring transformed this environment into a project-driven system of principal contractors and tiered sub-contractors, within which instability increases downwards through the layers.

The analysis of buyer-supplier relationships suggests that these changes have progressed to the extent that the embedded relationships between firms in the supply chain have withered and are in the process of dying. This atrophy is only too evident in the customer links of the surveyed firms which have become more price-based and arm's-length in character. At the same time, those same firms' supplier links are as socially embedded as ever, turning for the most part on relationships of trust and loyalty at the inter-firm level. Simultaneously, however, the labour market relationships of the smallest firms have been substantially ephemeralised, calling into existence employment agencies to maintain and facilitate labour flexibility.

What has appeared in this industry, therefore, is the progressive erosion and marketisation of inter-firm relationships downwards through the layers of the supply chain that supports the defence industries and, increasingly, the production of domestic/civilian electronics goods. The erratic, 'spiked' and unstable demand created by these processes, is eroding socially constructed economic relationships in the upper layers of the production chain. The camaraderie of small firms persists at the lower levels of the system, but at the expense of labour, which has been ephemeralised and marginalised.

It can be suggested, in conclusion, that the embedded, inter-firm relationships in the electronics industry, at least in the region covered by this study, are a relic of earlier times that are being eaten away by competition, corporate concentration, managerialism and the strict application of market-based principles by government. Whether the process that created small electronics firms in the region in the past will continue into the future is a moot point. It must be remembered, however, that the analysis presented here is based on only a small number of in-depth interviews. The study cannot pretend to produce definitive conclusions, only testable hypotheses that need to be matched against fuller economic data.

Note

1   The eleven firms interviewed were selected using a snowballing technique. The interviews were conducted in November and December, 1998, by Sarah Openshaw. Before the interviews, an effort was made to become conversant with the operations and terminology of firms in the sector to achieve open, detailed discussions (Healey

and Rawlinson 1993). Each interview lasted up to 90 minutes and was taped. The electronics firms interviewed can be separated into three main sectors: (1) printed circuit board (PCB) manufacturers (2 firms – PM1 and PM2); (2) printed circuit board assemblers who work with components and boards supplied by their customers (2 firms – PA1 and PA2); design and manufacturing firms who offer engineering services and solutions but have the facilities to undertake PCB assembly (7 firms – DM1 to DM7). A framework of question prompts was used to guide the interviews. These were based of the work of Uzzi (1996, 1997) and covered three broad areas. First they covered the background and foundation of the business, including ownership changes. Second, they explored business contacts with customers and suppliers and sub-contractors to gauge their nature, form and extent, and to explore issues of conflict and loyalty, and the impact of price and competition. Third, questions were asked on future relationships and change, to assess the longevity of relationships and the strength of external forces that have impacted or are impacting on the surveyed firms. It was possible for firms to exhibit embedded and arm's-length relationships at one and the same time. The interviews were transcribed verbatim. As far as possible the aim of the study has been to represent interviewees' views and comments in their own voice. It was recognised from the outset that the firms studies 'provide an especially fertile ground for embeddedness that might not exist for larger firms' (Uzzi 1997, p. 64). Therefore, the results are valid only for the electronics industry in the study area at the time of the interviews.

References

Asheim, B. and Isaksen, A. (2000), 'Localised knowledge, interactive learning and innovation: between regional networks and global corporations', in Vatne, E. and Taylor, M. (eds), *The Networked Firm in a Global World: Small Firms in New Environments*, Ashgate, Aldershot, pp. 163-198.

Bordieu, P. (1980), 'Le capital social: notes provisoires', *Actes de la Recherche en Sciences Sociales*, vol. 31, pp. 2-3.

Braczyk, H-J., Cooke, P. and Heidenreich, M. (eds.) (1998), *Regional Innovation Systems*, UCL Press, London and Bristol PA.

Chistensen, P. (2000), 'Challenges for Small Firm Sub-Contractors in an Era of Global Supply Chain Restructuring', in Vatne, E. and Taylor, M. (eds), *The Networked Firm in a Global World: Small Firms in New Environments*, Ashgate, Aldershot, pp. 67-92.

Coleman, J. (1988), 'The creation and destruction of social capital: implications for the law', *Notre Dame Journal of Law, Ethics and Public Policy*, vol. 3, 375-404.

Cooke, P. and Morgan, K.(1998), *The Associational Economy*, Oxford University Press, London.

Fedderke, J., Dekadt, R. and Luiz, J. (1999), 'Economic growth and social capital', *Theory and Society*, vol. 28(5), pp. 709-745.

Fukuyama, F. (1995), *Trust. The social virtue and the creation of prosperity*, Hamish Hamilton, London.

Healey, M. and Rawlinson, M. (1993), 'Interviewing business owners and managers: a review of methods and techniques', *Geoforum*, vol. 24(3), pp. 339-355.

Johannisson, B. (2000), 'Modernising the industrial district: rejuvenation or managerial colonisation?', in Vatne, E. and Taylor, M. (eds), *The Networked Firm in a Global World: Small Firms in New Environments*, Ashgate, Aldershot, pp. 283-308.

Lundvall, B.Å. and Johnson, B. (1994), 'The learning economy', *Journal of International Studies*, vol. 1(2), pp. 23-42.

Maskell, P., Eskilinen, H., Hannibalsson, I., Malmberg, A. and Vatne, E. (1998), *Competitiveness, Localized Learning and Regional Development – Specialization and Prosperity in Small Open Economies*, Routledge, London.

Maskell, P. and Malmberg, A. (1999), 'Localised learning and industrial competitiveness', *Cambridge Journal of Economics*, vol. 23, pp. 167-190.

Porter, M. (1990), *The Competitive Advantage of Nations*, Methuen, London.

Porter, M.E. (1998), *On Competition*, Macmillan, London.

Portes, A. and Sensenbrenner, J. (1993), 'Embeddedness and Immigration: Notes on the Social Determinants of Economic Action', *American Journal of Sociology*, vol. 98, no. 6, pp.1320-1350.

Putnam, R.D. (1993), *Making Democracy Work: Civic Traditions in Modern Italy'* Princeton University Press, Princeton NJ.

Rosa, P. and Scott, M. (1999), 'Entrepreneurial diversification, business cluster formation, and growth', *Environment and Planning C: Government and Policy*, vol. 17, pp. 527-547.

Staber, U. (1996), 'Accounting for differences in the performance of industrial districts', *International Journal of Urban and Regional Research*, vol. 20(2), pp. 299-316.

Storper, M. (1997), *The Regional World: Territorial Development in a Global Economy*, Guilford Press, New York and London.

Taylor, M. (1999) 'The small firm as a temporary coalition', *Entrepreneurship and Regional development*, vol. 11, pp. 1-19.

Taylor, M. and Conti, S. (eds) (1997), *Interdependent and Uneven Development: Global-Local Perspectives*, Ashgate, Aldershot.

Uzzi, B. (1996), 'The sources and consequences of embeddedness for the economic performance of organizations: the network effect', *American Sociological Review*, vol. 61, pp. 674-698.

Uzzi, B. (1997), 'Social structure and competition in interfirm networks: the paradox of embeddedness', *Administrative Science Quarterly*, vol. 42, pp. 35-67.

Woolcock, M. (1998), 'Social capital and economic development: towards a theoretical synthesis and policy framework', *Theory and Society*, vol. 27, pp. 151-208.

Chapter 7

Enterprise Embeddedness and Industrial Innovation in Spain: An Overview

Montserrat Pallares-Barbera

Introduction

Research on clustering and enterprise embeddedness in developed countries has been concerned mainly with economic activity in metropolitan areas. However, enterprise development and long-term economic growth also occurs away from these urban areas, in rural districts, based on the activities of small and medium sized enterprises (SMEs). The purpose of this chapter is to analyse the local area inter- and extra-firm networks of businesses in the non-metropolitan and rural districts in Spain that have been economically successful in the past decade, and to identify the role of embeddedness and innovation in shaping that success. Elements of the embeddedness model of local growth are clearly evident in these regions, and the overview presented in this chapter draws together and develops a critique of the evidence on the creation and success of inter-firm network structures in these non-metropolitan regions of Spain.

There is an extensive literature that suggests that inter-firm networks are central to enterprise innovation in local areas (Granovetter 1992, Grabher 1993, Braczyk *et al.* 1998, Taylor 2001). SMEs, especially family firms, often have limited capital to support research and development. Innovation among these firms is, therefore, dependent on the extra-firm networks to which they belong, that comprise other firms and both public and private institutions (Yeung 1994). Inter-firm collaboration and cooperation, based on trust, reciprocity and loyalty, creates embedded network structures that are inclusionary, persistent and relatively stable.

These social relationships between firms, reinforced by proximity, are argued to create long-term growth in local economies.

The argument presented in this chapter is divided into five sections. Following the introduction, Section 2 explores evidence on the nature and extent of innovation within non-metropolitan local production systems in Spain. Section 3 focuses on local firms' market strategies in these Spanish regions, and Section 4 explores the role and nature of institutional support in the processes of embeddedness. From this discussion, the concluding section poses a range of questions and develops a critique on embeddedness in the Spanish context, and the role that policy can play in shaping and stimulating these processes of growth and change.

Industrial Innovation and Local Production Systems in Spain

A range of processes can be identified as operating in non-metropolitan clusters in Spain that have operated to create their current success. The available evidence would suggest that there are four dimensions to these processes: initiation, promotion, network formation and innovation. In the following sections, each of these processes will be reviewed in turn.

The Initiation of Regional Industrialisation and Processes of Firm Formation

The existence of elements of an industrial tradition in a region is often seen as a source of positive externalities that produce spill-overs, supporting new network formation and reducing system uncertainty (Sabel 1987, 1989, Piore and Sabel 1986, Becatini 1988, Bartolini 1989, Bellandi 1989). Marshallian industrial atmosphere in most old industrial production systems was created even before the industrial revolution. In some Spanish regions, current manufacturing networks were initiated by proto-industrialisation dating back to the Middle Ages (Pérez 1999, Pallarés and Vera 2000). These initial proto-industrial networks created industrial cultures in those regions based on sectoral specialisation. Rueda's current industrial culture, for example, has its origins in the production of fine wines in that district (Sánchez *et al.* 1999).

Evidence would suggest, however, that in Spain in the 1990s, industrial tradition was not necessarily a basic factor stimulating growth and new firm formation in local economies. Some authors agree that industrial tradition generates tangible and intangible assets in local economies (Méndez *et al.* 1999, Pallarés and Vera 2000), but it does not operate as a propulsive mechanism in all districts. Alonso and Rodríguez (1999) argue, for example, that there was no tradition of a successful clothing and textile industry in Galicia before the last decades of the twentieth century, only tailors and seamstresses who worked to satisfy local demand without competition at the national level.

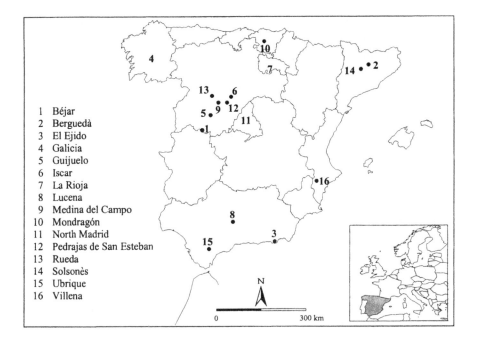

1 Béjar
2 Berguedà
3 El Ejido
4 Galicia
5 Guijuelo
6 Iscar
7 La Rioja
8 Lucena
9 Medina del Campo
10 Mondragón
11 North Madrid
12 Pedrajas de San Esteban
13 Rueda
14 Solsonès
15 Ubrique
16 Villena

Figure 7.1   Non-metropolitan clusters of embedded enterprise in Spain, 1999

Méndez *et al.* (1999, p. 226), however, suggest on the basis of evidence from Castilla-La Mancha, that 30 years of industrial tradition is enough to create strong local externalities and initiate cumulative growth in a region. Indeed, in some regions in Spain, industrial network structures were initiated as recently as the 1970s in, for instance, the textile and clothing district in Galicia (Alonso and Rodríguez 1999) (see Figure 7.1). In Villena, a similar industrial network was created in the footwear industries in the 1950s (Martínez and Alcazaras 1999).

Empirical research suggests that regional industrialisation develops and consolidates, in the Spanish context at least, when three sets of conditions prevail. First, it occurs when there is supra-local market expansion. Second, positive factor conditions in an area, such as a pool of labour or venture capital, may also promote industrialisation. And third, industrialisation can be initiated when endogenous firms stimulate imitative behaviour both inside and outside a particular region (Méndez *et al.* 1999, Pallarés and Vera 2000). This is what happened, for example, in the Galicia region, as international demand for its firms' textile and clothing products grew, and endogenous capital and know-how accumulated, attracting further investment from other parts of Spain (Alonso and Rodríguez 1999). Ondátegui (1999) drew similar conclusions from a study of metropolitan clusters in the Madrid region. He found that new firms were created in local production systems when existing firms' markets began to internationalise (even though they were unsure as to how to cope with international competition), when firms acquired new technological knowledge, when a pool of skilled labour developed, and when a good communication system existed.

Experience in Spain also suggests that local agricultural resources can also act as a foundation for the development of local production systems in non-metropolitan areas. In some cases, agriculture has been the foundation of value-adding chains, but only when other factors have been present. The wine sector in Rueda is an example of this process (Sánchez *et al.* 1999). The wine producing district of Rueda was on the point of collapse at the beginning of the 1990s. However, the combination of *verdejo* grapes, good soils, modernised wineries and intangible elements like tradition and local know-how brought growth to the district in the late 1990s fostered by cooperation. Equally, in El Ejido the modernisation of agriculture brought local growth (Silva 1999). Here growth is based on labour-intensive horticulture and greenhouse production within a coordinated agro-industrial supply chain of growers, food processors and distribution firms. This growth, however, has had its down side: dependence on cheap north African immigrant labour willing to tolerate low living standards; over-exploitation of aquifers together with soil salinity problems; problems caused by the extensive use of chemicals; the creation of a huge volume of vegetable waste; and the transformation of the landscape into a monotonous 'plastic sea' through the use of plastic greenhouses.

However, not all these conditions occur at the same time in all local production systems in Spain, and it is also clear that other factors are important in stimulating local growth. Local production systems in Spain are based principally on SMEs, endogenous capital and only few transnational plants (Méndez *et al.* 1999, Pallarés and Vera 2000, Caravaca 1999). Endogenous economic development in local regions in Spain is strongly affected by external processes, especially competition from newly industrialised countries (NICs), general market cycles and the processes of the global market.

Factors Promoting Growth in Local Systems

In most non-metropolitan local production systems in Spain, SMEs have been the engine of growth, and the modern business networks they contain were initiated in the 1960s and 1970s. Since then, the growth and decline of those regions has paralleled the success of the local manufacturing sector. Most of the regions have a lead sector, for example the wood sector in Lucena (Pérez 1999) and the textile sector in Berguedà (Pallarés and Vera 2000). However, recession in the early 1990s brought the restructuring of these older lead sectors, forcing diversification. In Lucena, restructuring saw the emergence of graphics and industrial refrigeration, and in Berguedà the emergence of metal working and food industries. As a result, while these districts lost population at the beginning of the 1990s, by the end of the decade they had high rates of regional economic activity, increasing occupation rates, low unemployment rates, and rising rates of GDP per capita.

The Formation of Networks of Relationships in Local Systems

Evidence would suggest that the growth of local economies in Spain is based on the creation of enterprise networks built on trust. These networks give cohesion to a district's sectors, cement ties with external markets, guarantee access to product innovation, and foster competition. Local production systems in Spain also appear to have a number of distinguishing characteristics. First, the regions within which these systems occur tend to have a medium-sized city at their heart that supplies urban services and dominates a local hierarchy of settlements. Second, the success of local production systems is based on endogenous growth, with the capital generated by local entrepreneurs being reinvested in the system. As part of this process of growth, entrepreneurs improve the quality of their designs, increase capacity and exercise control over distribution. At the same time, they create extensive subcontracting networks (Alonso and Rodríguez 1999). Third, personal relationships between entrepreneurs capitalise on experience and foster local knowledge flows. Fourth, the commercial success of local entrepreneurs creates a local feeling of pride that may even spread to the labour force. In Lucena, for example, it has been reported that labour is willing to work extended hours and to accept reduced vacation time. Entrepreneurs in these localities express their local pride by consuming locally, and the regional feeling that is engendered has been summarised in the phrase, 'We are eating the national market' (Pérez 1999, p. 45).

The processes creating success have not been the same in all regions. In some places, supply side processes have been important, especially the relocation of mature, labour intensive and energy intensive industries seeking lower labour costs (Méndez *et al.* 1999, Caravaca 1999, Alonso and Rodríguez 1999). New labour-intensive ventures created in this way – by relocation, as joint ventures or green field investments – tend to create ephemeral jobs with precarious labour contracts, supported by home-working (González 1999). The local production system in Pedrajas de San Esteban, for example, is based on firms using local female labour, without any element of innovation (Sánchez *et al.* 1999). On the demand side, firms may establish contracts among themselves to buy in supplies for the local production chain, but these arrangements do not occur in all local production systems. Indeed, relationships beyond local production networks may be important to stimulate product and process innovation and R&D, an innovation in management (Hernández 1999).

Territorially, successful local production systems also appear to need local institutional support, for example through the provision of industrial space and technology centres, and in a local cohort of both similar firms and sectorally diverse firms (Mecha 1999).

Innovation Processes in Local Production Systems

New and dynamic firms in a region generate positive externalities and create an innovative milieu within a local production system that is attractive to new firms (Pérez 1999). Innovations in products, process, logistics and technology, when combined with flexibility, create comparative advantage and global competitiveness in a local economy (Pallarés and Vera 2000). This innovation can be both defensive (to meet new market demands) and offensive (to create new market conditions). The combination of these different strategies has been recognised in the success of the La Rioja region by Climent (1999). Here, innovation in the machinery used in production increased product quality. It also increased production and introduced new products to an educated global market. Innovations are, in fact, a combination of the knowledge of the entrepreneur and, the interaction between entrepreneurs and clients (Ondátegui 1999).

Local branding, through the use of a trademark, can also help a local production system penetrate international markets by promoting the idea of good quality, innovative design and creativity. This is a strategy that has been adopted successfully in Rueda in relation to the production of fine wines (*Denominación de Orígen Rueda*) (Sánchez *et al.* 1999), in Ubrique for the marketing of leather products (González 1999), and in Galicia to promote textiles and clothing products (*La Moda Gallega*) (Alonso and Rodríguez 1999).

Exogenous investment can also revitalise older firms in local production systems through joint ventures or by setting up green field enterprises. In Rueda, for example, exogenous and endogenous investment combined to create their own wineries and to vertically integrate vineyards and the wine industry in the region. Unlike transnational investment, this type of exogenous investment becomes closely involved in the local production system of a region.

Improvements in telecommunications technology have reduced the impact of distance on the operation of local production systems, though they have not altogether eliminated the constraints it imposes on the physical flows of goods (Ondátegui 1999). Thus, Spanish local production systems need to be reasonably well connected to their most important local and national markets in order to be successful. Historically, a lack of roads has left some regions isolated, though recent public investment has gone some way to alleviating this problem. Thus, Sánchez *et al.* (1999) claim that Rueda's relative position on the Castilla-Léon axes of communication, that connect Rueda with the Castillian industrial belt and with Madrid and La Coruña, is one of the basic pillars of the region's production innovation.

The Market Strategies of Firms in the Spanish Regions

Entering international markets and gaining an international reputation are essential processes creating the innovative milieu and inter-firm network of a local production system. Most firms within Spain's local production systems sell on international markets and are competitive by international standards (Mecha 1999). The Rueda region, for example, which historically sold wine to local and national markets, has produced a low grade white wine that is competitive on international markets, principally in Germany and Italy (Sánchez *et al.* 1999). However, most specialised local production systems in Spain compete on national and international markets on quality and design, and not on low cost. Galicia's textiles and clothing output is aimed at medium to high quality export markets (Alonso and Rodríguez 1999), and Rueda produces fine wines (Sánchez *et al.* 1999). Sophisticated national demand in Spain has driven up product quality, presentation and distribution, so they now meet international standards (Alonso *et al.* 1999)

In contrast, some SMEs in local production systems are tied into major distribution chains. This arrangement offers both advantages and disadvantages. Certainly, it provides the firm with steady and secure demand and helps to achieve a level of quality control. However, excessive dependency is a threat to the firm because its competitive strength is low cost production rather than a brand name, which reduces added value (Méndez *et al.* 1999). It is the buyers who tend to control these supply chains, reducing the firm's discretionary power over things like transport (Mecha 1999). What is more, the firm can have no independent market strategy, increasing its vulnerability. The wood and furniture industry of Medina del Campo suffers all these distribution chain problems (Sánchez *et al.* 1999).

Rivalry between local firms can stimulate innovation and improvement in the specialised products of a region. Exogenous investment can also have the same effect, demonstrating alternative strategies for production and competition. Thus, Marqués de Riscal, an outside investor that moved into the Rueda region, introduced a sauvignon grape that has become the foundation of the local wine industry's international competitiveness. Smaller wineries have imitated the company's approach and have renovated their production facilities. This has allowed them to produce fine wines themselves and to meet the quality demanded in international markets (Sánchez *et al.* 1999).

Institutional Support and Enterprise Embeddedness in Spain

The consolidation and continuing success of local production systems in Spain has also been shown to require in some circumstances the support of local and non-local institutions from both the public and the private sectors. These institutions have the potential to promote collaborative and cooperative relationships, based on trust, reciprocity and loyalty, and to stimulate processes of network formation in incipient local economic clusters (Mecha 1999). Research in Spain has recognised a range of agencies that promote local economic development that are public as well as private and supra-local as well as local (Méndez *et al.* 1999). In some regions, these institutional mechanisms have been active in generating innovation, creating trademarks, promoting cooperation and developing a cordial climate among networked firms. In wine production in Rueda, firms have received a lot of technological support and help with commercial distribution from public institutions (Sánchez *et al.* 1999). Indeed, a number of researchers have concluded that the absence of these institutions and the support they offer is characteristic of the less developed parts of Spain (Pallarès and Amorós 2000, Pallarés and Vera 2000).

Local culture is at the heart of this supportive tissue of institutions in some Spanish regions. In Lucena, for example, the social relationships of the local cluster that has been built on the wood and furniture sectors are cemented and reinforced by the processions (*procesiones*) that are organised during Easter each year (Pérez 1999). This local cultural event is organised by local entrepreneurs and their workers, and serves to consolidate and strengthen human and social ties that cut across the hierarchies of owner and worker. This identification of people with a place is also evident in the economic development of the Mondragón region, focused on the locally deep-rooted Basque culture and its regional language. In this particular instance, the impact of local culture was magnified by the activities of a local priest who strove to develop the capabilities of local people, leading to the creation of cooperative ventures and enduring economic projects (Torres 1999).

In contrast, there are successful local production systems in Spain that have no such local supportive institutional tissue (Sánchez *et al.* 1999). In the leather industries of Ubrique in Andalucia, for example, there is flexible specialisation and strong individualism and only incipient support through local social and institutional agents (González 1999). In this district, cooperation between firms is hard to achieve even though they are linked through outsourcing and subcontracting strategies within the local value-added chain. Competition is strong in Ubrique over innovation in materials and in product design, and this has built a local entrepreneurial spirit that has had significant positive spin-off effects.

Individualism and competition also typify inter-firm relationships among the textile firms of Bejar. Cohesion in the cluster in this district derives only from the necessities of the local technical division of labour, and the first signs of cooperation only appeared in the district in the later 1990s (Sánchez *et al.* 1999). What is more, the institutional support of government, in the form of economic incentives, has not yet managed to expand the markets for this district's products into the international arena. In the footwear district of Villena, it has been the process of spill-over from neighbouring regions, without institutional support, that has created success. This success has combined the manual skills of 'grupicos' – small groups of three workers who had acquired their skills in neighbouring regions' footwear firms ('Valencianas') – and 'capitalist' partners from agriculture, wine distribution and banking, who had access to finance and expertise in marketing (Martínez and Alcaraz 1999). Individualism has similarly underpinned textile and clothing development in Galicia, but without inter-firm cooperation and institutional support.

Evidence would suggest, moreover, that firms in Spain's non-metropolitan industrial districts do want institutional help, but of a very specific type. Principally, they want government assistance with the one factor that most often holds them back – a shortage of skilled labour. This was the case in Berguedà at the end of the 1990s when entrepreneurs asked local educational institutions to support them by establishing specialised training courses (Pallarés and Vera 2000). In other situations, the role demanded of local institutions has been to assist firms to improve research, to increase the quality of production. This, for example, was the demand placed on the Estación Enológica de Castilla y León, to provide direction on grape harvesting and quality control of fine wines.

Conclusions

From the discussion of this chapter it is clear that the embeddedness concept, as the basis for local economic growth and clustering, is an elusive idea. The examples from Spain identify a range of factors that might promote economic growth and development of local production systems in non-metropolitan areas, though there is no consistency from example to example: no clear identification of what conditions are necessary and/or sufficient to initiate and sustain that local growth. In a classic sense, the embeddedness model of local economic growth and cluster formation is built on a set of stylised facts that raise more questions than they answer. Some researchers stress the roles of innovation and collective learning. Others emphasise tradition and collaborative network practices. Moreover, as stylised facts, what is 'innovation' and how should 'tradition' be interpreted? When does cooperation supplant competition, and what are the limits of 'trust'? Indeed, how do learning mechanisms arise and persist?

From the present analysis, three overarching questions arise concerning the propositions of the embeddedness model. First, how are embedded relations initiated in a local production system? In this context, is industrial tradition a necessary and sufficient condition? What roles are played by sectoral specialisation, specialised local labour pools, locally available venture capital, and market expansion, for example? A fundamental problem of the model is that it is a-temporal and elaborates processes only after they have been initiated. Second, how is innovation created and spread through local production systems? Is it simply through processes of circular and cumulative causation (Myrdal 1956, Friedman 1972), and how do these processes match with the needs and strategies of individual firms? Indeed, how do embedded processes of innovation marry with the processes of spatial uneven development that are basic to the capitalist system? Third, what mix of factors promotes local growth? Success in non-metropolitan areas appears to relate to the operations of SMEs, endogenous capital, sectoral specialisation, and medium-sized cities as hubs of relational flows (financial, political and social), and social ties between entrepreneurs and labour within districts (including local pride). It is difficult to identify 'one best way'.

Nevertheless, the present analysis of non-metropolitan local production systems in Spain suggests that there may be a significant role to be played in their initiation and promotion by the institutional support of government, in the shape of policies and programmes to promote local economic development (see Alonso *et al.* 1999). On the basis of the empirical studies on Spain reported in this chapter, it can be suggested that two levels of policy might be appropriate to achieve these local growth goals. First, at the firm level, there is the need for financial support, and programmes to build the skills of entrepreneurs and their workers (Pallarès and Amorós 2000). Such policies also need the support of databases of information on competitors, markets, technology and available public incentives, as well as the creation of local enterprise 'incubators' to create new firms. Second, at the network level, it would appear that programmes can be usefully used to promote group identities through participation in fairs and specialist exhibitions (Méndez *et al.* 1999), and support regional technology centres. They can be used to promote innovation programmes, and also to promote regional urban centres to strengthen local urban demand for a production system's output.

This analysis of embedded inter-firm relationships in non-metropolitan Spain offers clear support for the processes that drive the embeddedness model of local growth. With the right combination of local circumstances, in terms of resources, entrepreneurships, labour supply, networking and social cohesion, it would appear that clustering and embeddedness can bring significant and dynamic growth to non-metropolitan regions. Also, a supportive tissue of institutions has also been important in promoting and shaping that growth.

References

Alonso, M.P. and Rodríguez, R. (1999), 'La industria de la moda en Galicia. Innovación y desarrollo local en una pequeña ciudad gallega: Lalín', *Proceedings of the II Seminar on Innovación Industrial y Desarrollo Local en la Península Ibérica*, Salamanca, Grupo de Geografía Industrial, AGE, pp. 3-22.

Alonso, J.L., Aparicio, J. and Amador, J.A. (1999), 'Factores territoriales y procesos de innovación en los sistemas productivos locales de Castilla y León". *Proceedings of the II Seminar on Innovación Industrial y Desarrollo Local en la Península Ibérica*, Salamanca, Grupo de Geografía Industrial, AGE, pp. 23-38.

Bartolini, S. (1989), *Elementi di una Microfondazione della Teoria dei Distretti Industriali*, Paper presented at the Barcelona Institute of Metropolitan Studies, Bellaterra.

Becatini, G. (1988), *Some Thoughts on Marshallian Industrial Districts. A Socioeconomic Notion*, Paper presented at the Barcelona Institute of Metropolitan Studies, Bellaterra.

Bellandi, M. (1989), 'The industrial district in Marshall', in E. Goodman and J. Bamford (eds), *Small Firms and Industrial Districts in Italy*, Routledge, London, pp. 136-152.

Braczyk, H.J., Cooke, P. and Heidenreich, M. (eds) (1998), *Regional Innovation Systems. The Role of Governance in a Globalized World*, UCL Press. London.

Caravaca, I. (1999), 'Sistemas productivos y medios innovadores en Andalucía', *Proceedings of the II Seminar on Innovación Industrial y Desarrollo Local en la Península Ibérica*, Salamanca, Grupo de Geografía Industrial, AGE, pp. 39-82.

Climent, E. (1999), 'Sistemas productivos locales en la Rioja. Innovación ante el reto de la globalización económica', *Proceedings of the II Seminar on Innovación Industrial y Desarrollo Local en la Península Ibérica*, Salamanca, Grupo de Geografía Industrial, AGE, pp. 83-116.

Friedman, J. (1972), 'A general theory of polarised development', in N.M. Hansen (ed), *Growth Centres in Regional Economic Development*, Free Press, New York, pp. 82-107.

González, G. (1999), 'Sistemas productivos locales y medios innovadores en Andalucía. El caso de Ubrique', *Proceedings of the II Seminar on Innovación Industrial y Desarrollo Local en la Península Ibérica*, Salamanca, Grupo de Geografía Industrial, AGE, pp. 117-150

Grabher, G. (ed) (1993), *The Embedded Firm. On the Socioeconomics of Industrial Networks*, Routledge, New York.

Granovetter, M. (1992), 'Economic institutions as social constructions: A framework for analysis', *Acta Sociologica*, vol. 35, pp. 3-11.

Hernández, S. (1999), 'Las peculiaridades de los espacios de innovación en los territorio insulares: El caso de Canarias', *Proceedings of the II Seminar on Innovación Industrial y Desarrollo Local en la Península Ibérica*, Salamanca, Grupo de Geografía Industrial, AGE, pp. 151-177.

Martínez, A. and Alcaraz, R.S. (1999), 'La industria en la comarca del Alto Vinalopó (Alicante). El caso del sector del calzado entre la tradición y la innovación', *Proceedings of the II Seminar on Innovación Industrial y Desarrollo Local en la Península Ibérica*, Salamanca, Grupo de Geografía Industrial, AGE, pp. 1-25.

Mecha, R. (1999), "Sonseca: Un estudio de caso de incipiente medio innovador', *Proceedings of the II Seminar on Innovación Industrial y Desarrollo Local en la Península Ibérica*, Salamanca, Grupo de Geografía Industrial, AGE, pp. 177-202.

Méndez, R., Rodríguez, J. and Mecha, R. (1999), 'Medios de innovación y desarrollo local en Castilla-La Mancha', *Proceedings of the II Seminar on Innovación Industrial y*

Desarrollo Local en la Península Ibérica, Salamanca, Grupo de Geografía Industrial, AGE, pp. 203-234.

Myrdal, G. (1956). *Economic Theory and Underdeveloped Regions*, Methuen, London.

Ondátegui, J. (1999), 'Territorio e innovación en el norte metropolitano de Madrid', *Proceedings of the II Seminar on Innovación Industrial y Desarrollo Local en la Península Ibérica*, Salamanca, Grupo de Geografía Industrial, AGE, pp. 235-260.

Pallarès, M. and Amorós, J. (2000), *La Indústria del Solsonès. Situació Actual i Valoració de Potencial*, Consell Comarcal del Solsonès, Solsona.

Pallarés, M. and A. Vera (2000), 'Incrustación Industrial y Medio Innovador en la Comarca del Bergadà', in J.L. Alonso and R. Méndez (eds), *Innovación, pequeña empresa y desarrollo local en España*, Civitas, Madrid.

Pérez, B. (1999), 'Sistemas productivos locales y medios innovadores en Andalucía. El caso de Lucena', *Proceedings of the II Seminar on Innovación Industrial y Desarrollo Local en la Península Ibérica*, Salamanca, Grupo de Geografía Industrial, AGE, pp. 275-312.

Piore, M. and Sabel, C. (1986), *The Second Industrial Divide*, Basic Books, New York.

Sabel, C. (1987), *The Re-emergence of Regional Economies*, unpublished paper.

Sabel, C. (1989), 'Flexible specialisation and the re-emergence of regional economies', in P. Hirst and J. Zeitlin (eds), *Reversing Industrial Decline*, Pergamon, Oxford, pp. 17-70.

Sánchez, J.L., Aparicio, J. and Alonso, J. (1999), 'Procesos de innovación en industrias rurales tradicionales: La elaboración de vino en la denominación de orígen Rueda', *Proceedings of the II Seminar on Innovación Industrial y Desarrollo Local en la Península Ibérica*, Salamanca, Grupo de Geografía Industrial, AGE, pp. 313-332.

Silva, R. (1999), 'Sistemas productivos locales y medios innovadores en Andalucía. El caso de El Ejido', *Proceedings of the II Seminar on Innovación Industrial y Desarrollo Local en la Península Ibérica*, Salamanca, Grupo de Geografía Industrial, AGE, pp. 333-368.

Taylor, M. (2001), 'Enterprise, embeddedness and local growth: inclusion, exclusion and social capital', in D. Felsenstein and M. Taylor (eds), *Promoting Local Growth: Process, Practice and Policy*, Ashgate, Aldershot, pp. 11-28.

Torres, K. (1999), 'El medio innovador de Mondragón', *Proceedings of the II Seminar on Innovación Industrial y Desarrollo Local en la Península Ibérica*, Salamanca, Grupo de Geografía Industrial, AGE, pp. 333-368.

Yeung, H. W. (1994), 'Critical reviews of geographical perspectives on business organisations and the organisation of production: Towards a network approach', *Progress in Human Geography*, vol. 4, pp. 460-490.

Chapter 8

Local Embeddedness in Global Financial Services: Australian Evidence on 'The End of Geography'

Pierre Agnes

Introduction

This chapter contributes to debates on globalisation and local embeddedness by discussing these interrelationships in the context of financial services. Some accounts of financial services restructuring argue that globalisation has led to the 'end of geography', denoting the declining importance of spatial processes in patterns of financial production. In this interpretation, electronic technologies, innovation, financial market deregulation, and multinational bank conglomerates are the protagonists of hyper-mobile finance.

Despite these globalisation processes, the network of world financial centres that serve as the hubs of the financial services industry still persists, reflecting the importance of local embeddedness in systems of financial services production. Theoretical frameworks in the geography of financial services suggest that local embeddedness stems from the creation of dynamic economies. These are characterised by flows of information in formal and informal networks among rival banks and business services firms, local financial market characteristics and conventions, and the existence of skilled finance labour.

There is comparatively little empirical evidence of these processes of embeddedness in wholesale financial services, and this is problematic because much research has stressed the prevalence of this model in other economic sectors. The financial services industry is comprised of many sectors with different geographies, and this chapter uses data from qualitative interviews with finance professionals in Australia to explore different patterns of local embeddedness in three global financial services: swaps dealing, futures broking, and master custody.

The 'End of Geography' in Financial Services? Global Banks and Local Embeddedness in World Financial Districts

In the context of financial services, the relationships between globalisation and local embeddedness have been framed within the 'end of geography' debate (O'Brien 1992). This metaphor ascribes dominance to the potential of several interrelated restructuring processes in reducing the importance of local embeddedness in world financial centres. Electronic trading and funds transfer systems have facilitated the international movement of money, and lessened the ability of domestic financial market regulations to achieve their desired aims. This provided the rationale for many OECD nations to deregulate their financial markets during the 1980s (Thrift and Leyshon 1994). Although the nature and extent of liberalisation varies and remains uneven internationally, nations have dismantled anti-competitive regulations, removed restrictions on financial transactions, and opened domestic capital markets to foreign investors and borrowers. Deregulation combined with electronic technologies, has propelled securitisation, financial innovation, and the proliferation of many new off-balance sheet instruments. These developments have contributed to financial market integration and the expansion of 24-hour trading (Thrift and Leyshon 1988, 1994).

Globalisation has also been accompanied by the formation of multinational financial services conglomerates as the central organisers of the global financial system (OECD 1993, Thrift 1994, Pryke and Lee 1995). These conglomerates emerged through national and international mergers and acquisitions that have accelerated in the past decade. The quest for larger size and economies of scale and scope are perceived as necessary prerequisites to successfully compete in the global financial environment (Llewellyn 1992). This is related primarily to two factors. The first is the concern for credit risk, and the imposition of capital adequacy standards by the Bank for International Settlements, that have increased the cost of capital for banks and influenced the competitive position of financial conglomerates. Credit rating has become the paramount consideration in derivatives and other lines of business that are risky. The second factor is the goal of many conglomerates to form a globally integrated full-service securities business to provide 'one-stop shopping' for clients, especially institutional investors. This has led to progressive diversification in product and geographic markets (Thrift and Leyshon 1988). In part, globalisation by financial conglomerates is a strategy to service clients abroad to prevent loss of domestic market share to foreign competitors (Kindleberger 1983). These changes have contributed to the concentration of business whereby a few global banks dominate trading volumes in many financial services.

Despite globalisation, there has not been a decline in the role of spatial factors in the organisation of financial services production. Global restructuring processes have led to much greater specialisation, diversity, and differentiation in the international system of financial centres than typical hierarchical classifications suggest (Daniels 1993). Furthermore, within financial cities multinational banks locate in close spatial proximity to form specialised financial districts (Daniels 1986, Gad 1991, Moss and Brion 1991). Global banks locally-embed themselves in financial districts for the following reasons:

• financial markets are based on the rapid dissemination and exchange of information in informal networks;
• financial markets are based on formal business networks and market conventions;
• networks between banks and corporate and institutional clients;
• localised skilled finance labour.

Embeddedness Due to the Exchange of Financial Information in Informal Networks

Financial services are based on the collection and synthesis of market information that enables them to advise clients in their investment and management decisions, and the acquisition of market intelligence is an essential dimension of financial services. Considerable costs are embodied in the provision of financial services and in searching for specialised information suppliers and potential collaborators. The process of information gathering increases pressure for local embeddedness (Thrift 1987, 1994, Peet 1992, O'hUallachain 1994).

Financial services have different informational contents with implications for local embeddedness (Clark and O'Connor 1997).

Translucent products are those where the general principles are universally known, yet their production relies on highly localised sources of informal information, knowledge of market practices, and specialist expertise. In contrast, *transparent products* are highly global and homogeneous. Finally, *opaque products* are highly specialised unique transactions that are extremely localised in nature.

Given the volatile nature of many financial markets, timely access to information is often essential because it has a limited lifespan in terms of its utility. The same information may have multiple meanings and ramifications for different banks and traders due to their market positions and business goals. In some cases the information may be vague, ambiguous, and difficult to interpret and understand.

However, the gathering of market information is only an initial step in the production of financial services. Market information is distinctive from market intelligence. Market information includes both standardised routine information (e.g. prices of commodities, quantities for sale or purchase, economic information on interest rate movements and unemployment figures), and specialised non-routine information (e.g. specific legal advice in securities contracts, auditing). This is compounded by the increasing specificity and lack of homogeneity of information, and the increasing fragmentation of knowledge where specialists have single pieces of information (Tornqvist 1983). Market information only acquires value when it is analysed and several pieces of information are assembled and transformed into market intelligence, so that finance professionals are able to develop a 'market view' that has usable and tradable value.

This process of valorisation occurs primarily through informal information trading networks despite the growth of electronic communication media (Ter Hart and Piersma 1990, Thrift and Leyshon 1994, Thrift 1994, Pryke and Lee 1995). Much information is characterised by its complexity and embodiment in individuals. This 'tacit' information is obtained and transmitted through face-to-face contacts in order to clarify ambiguity through the perspectives of collaborators and rivals. Face-to-face contact provides opportunities to make judgements about the quality of information and expertise of potential information partners in particular situations. Many contacts are unplanned and occur simply by 'being there'. Interpersonal networks are characterised by cooperation, reciprocity, trust, and reputation (Abolafia and Biggart 1991, Amin and Thrift 1993, Grabher 1993). Synergies accumulate in networks, and firms are not independent actors but are parts of complex webs of interfirm relationships. These social features are used in formal business transactions that are difficult to quantify and would not occur without prior knowledge of a counterpart's trustworthiness due to the risks of failure and opportunism.

Embeddedness for Access to Place-Specific Trading Networks and Market Conventions

The production of financial services occurs in dynamic and flexible formal networks of interlinked rival financial firms and supplier firms that create, manipulate, and exchange the information discussed above (Ter Hart and Piersma 1990, Thrift and Leyshon 1994, Thrift 1994, Pryke and Lee 1995).

The types of formal networks vary depending on the specific financial service. *Funds management* requires frequent interpersonal contacts between investment analysts and the companies they monitor and invest in. *Interbank activities* involve banks traded *directly* with dealers in rival banks. Some financial services are traded on open-outcry floors as in the case of some futures markets; many other financial services occur in over-the-counter markets formed by the telecommunication and computer links among banks, as in foreign exchange and swaps. *Brokered markets* are those that have brokers as intermediaries to find trading counterparts in transactions. In any case, interbank networks are characterised by a competitive and cooperative paradox: banks are often competitors, but this coexists with cooperation (Ter Hart and Piersma 1990). Local embeddedness is important because although interbank transactions occur by telephone or Reuters direct dealing system, dealers prefer to work with rivals they know and through an established relationship.

Other types of financial services require a range of financial and other specialist business services firms in accounting, securities and banking law, and computing. Mergers and acquisitions advising, for example, requires the interaction among commercial and investment banks, securities underwriters, and professional services in law, accounting, and insurance (Thrift 1987, O'hUallachain 1994). When complex deals are assembled, it is important that clients, banks, and business professionals are locally embedded. Similarly, Daniels' (1986) research revealed that the choice of London and New York as locations for foreign banking operations is highly influenced by the existence of agglomeration economies and outsourcing of some services (insurance, law, real estate) to external firms.

Formal banking networks are also characterised by the *mutual orientation* of counterparts, referring to the common knowledge, language, and rules in relation to technical matters and procedures, and unwritten social norms and customs particular to that industry (Amin and Thrift 1993, Thrift 1994). Mutual orientation arises from a common industry work culture that is easier to reproduce by the spatial clustering tendency of professionals. In addition, mutual orientation also arises from industry associations and training institutions that coordinate information flows, skills and technological development, and industry attitudes.

Embeddedness for Client Relationship Management

Local embeddedness is important in financial services due to their relationship-based client-orientation (Ter Hart and Piersma 1990, Thrift 1994). Relationship management has grown in importance during the past decade with the proliferation of customised off-balance sheet instruments designed to suit the exact needs of clients. Clients are not just passive end-users of services, they are integrated as central elements in the production process by providing firm-specific information. Most client-service firm interaction is characterised by *sparring*: transactions are interpersonal and reciprocal, and professionals and clients guide each other (Tordoir 1994). Close client relationships and relationship management have grown in importance during the past decade with the proliferation of customised financial innovations, such as swaps, which are designed to suit the precise needs of the clients. Some clients have specific goals and clear ideas about their requirements, while others may be completely dependent on the service-provider for advice and expertise (Aharoni 1993). Customised and complex high value-added transactions require firm-specific and specialised information, greater information processing, extensive communication, and negotiation that occur through face-to-face meetings. Some research argues that clients are more transactional, thus leading to the decline of trust and intensifying the importance of relationship management through embeddedness (Thrift 1994).

Embeddedness to Access Skilled Local Labour Markets

The production of financial services is labour-intensive, and requires highly skilled finance personnel that possess technical knowledge and skills. The abilities of banks to advise clients in their investment and financial management decisions are based on the accumulated knowledge, reliability, responsiveness to clients, and reputation of banks as major strategic assets. These abilities are embodied in highly educated and multi-skilled specialist finance personnel who work in teams that understand how to apply information and knowledge flexibly to solve a range of problems (Bertrand and Noyelle 1988, Storper and Scott 1990, O'hUallachain 1994). Financial labour markets are comprised of highly skilled *front office* banking professionals that command high salaries based on their abilities to perform sophisticated and complex functions. In addition, the availability of a lower-paid, but often equally skilled local pool of *back office* accounting, clerical, and settlements personnel also contributes to the local embeddedness of financial firms. For financial personnel, embeddedness is important for their career development because financial services labour markets are relatively closed with job vacancies advertised by word-of-mouth through social networks. Local knowledge of prospective personnel and their capabilities is very important (Peet 1992).

A substantial weakness with existing studies of financial services is the assumption that these processes contribute to the local embeddedness of banks in world financial centres, despite the lack of empirical research. In other industries, research has demonstrated that there is a complex interplay between globalisation and local embeddedness that contributes to qualitatively different manifestations of territorialisation along a continuum (Storper 1997, Markusen 1996). The next section contributes to this globalisation and local embeddedness debate by comparing embeddedness in three financial services: swaps dealing, futures broking, and master custody.

The Roles of 'Space', 'Place', and 'Embeddedness' in Financial Service Transactions

Swaps dealing, futures broking, and master custody are three different sides of the financial services industry. Swaps and futures are part of the burgeoning derivatives industries, and illustrate front office trading services and the process of financial securitisation that has expanded tremendously since the 1980s. Both are financial instruments similar to insurance and allow corporations, institutional investors, and government agencies to protect their investments from losses associated with volatility in foreign exchange and interest rates (Napoli 1992, Daugaard and Valentine 1995).

Whereas swaps are customised transactions in over-the-counter markets, formed by electronic spaces of dealing rooms of banks, futures are contracts with standard specifications (i.e. their size, expiry and settlement dates, values) are purchased and sold on exchanges by member broking firms. On most exchanges, trading occurs electronically on screen-based systems; most U.S. exchanges still trade by open outcry on a physical floor.

In contrast, master custody involves *back office* data processing services associated with the investment process. Pension funds, mutual funds and other collective investment vehicles, and insurance invest funds in a range of securities markets worldwide. This combination of greater scope and scale of investment opportunities has made the investment process more sophisticated. Investment managers outsource these functions to *master custodians*, who process and synthesise investment data supplied by their *sub-custodians* in domestic markets, and provide consolidated accounting reports, and other value-added services like portfolio evaluation, performance measurement, and derivatives reporting.

Embeddedness in Interest Rate Swaps Dealing

> You cannot run an Aussie dollar swaps book anywhere effectively but in Sydney. You need pure market information, talking to people, and need to see all the information flows coming through to do all the pricing accurately. If I wasn't here, I don't think I'd get the same information even with a voice box. I don't think I'd have the same *'feel for the market'*, for what is happening to spreads and why. I don't think I'd be as good friends as I am with a number of people and therefore I wouldn't get the same information flows. You wouldn't be a part of that network if you're located elsewhere. (Swaps Dealer, British Bank)

Although organised primarily by multinational banks (MNBs), the swaps industry epitomises the processes of local embeddedness. As the quote above reveals, swap banks are locally embedded in financial districts due to the importance of information flows in dealing networks comprising rivals, swaps and futures brokers, and clients. The dealing orientation means that swap desks operate as independent profit centres and determine their own business directions. The only manifestations of globalisation are related to centralised reporting of risk positions so the parent bank can determine its aggregate global risk exposures. Swaps are provided through intra-corporate networks that interlink localised swap markets and provide access to global markets from a local base.

Whereas firms in other sectors seek to eliminate risk through externalising production functions in inter-firm networks, swap banks are in the business of warehousing swaps in a portfolio and managing risks to earn profits (Casserley 1991). This is done in formal hedging networks with rival swap banks, and swaps and futures brokers. The important feature of swaps dealing is that every transaction can be managed in numerous ways. For example, a 10-year swap can be broken down and managed using combinations of one, three, and five-year swaps as mutual offsets; hedged internally within the banks' swaps book; using interest rate futures on the Sydney Futures Exchange; or with elements of all. Likewise, parts of the swap could be warehoused for periods of time, while other portions are hedged immediately. Although the configuration of swap networks depends on numerous factors (e.g. client-specific information on the size and purpose of the swap, the position and shape of the banks' swap portfolio, whether the swaps market is rallying, the volume of business going through the market, etc.), dealing decisions are influenced by subjective considerations. For example, dealers use swaps brokers because:

> When markets are volatile it is easier to use a broker rather than dealing direct because you can get the deal done quickly rather than having to ring around and find out where the other side of the deal and your counterpart might be. At other times, anonymity is preferable, so you'll deal direct. (Swaps Dealer, British Bank)

Similarly, several futures firms are used to camouflage business and prevent rivals from observing their business:

> There's a good reason for using other brokers. The swaps market is quite small and there are few secrets out there – if people know a large deal has gone off, they'll know how many futures must be sold and they'll move against you to earn big money! (Futures Dealer)

A related factor is that swap markets remain organised on the basis of national currencies, and market conventions require swap banks to be locally embedded. A dealer explained that:

The Australian market is characterised by illiquidity in some segments and inefficiencies in others. The rate resetting in short-term swaps differs from elsewhere. Short-term risk is more difficult to manage in Australia than say in the US due to the bank bill market and its rates. (Swaps Dealer, British Bank)

The importance of information acquisition also contributes to the local embeddedness of swap banks. Dealing networks are essential conduits for the transmission of market information, and are characterised by flexibility, dynamism, and multiplicity of relationships with numerous market players to facilitate rapid exchange of information. Of greatest significance is the informal trading of tacit business information underpinning formal networks. These informal relationships provide dealers with information interpretation that enables them to develop a 'market feel' that shapes bank-dealing decisions. Consequently, information-trading networks are characterised by trust, confidence, cooperation, and reciprocity.

Social features are important because although dealers are wary of giving information to rivals, futures or swap brokers, their dealings in the market reveal strategic business information. This causes concern because brokers deal with many rival banks and are privy to specific information about the market activities of dealers. This requires trust and confidence. As one dealer remarked:

Trust is important – especially on the point of discretion. You need to know that you can trust brokers and their judgement that if you need to have discretion in executing an order, they will work the order to your advantage. You want information about who's doing what, how aggressively. (Swaps Dealer, American Bank)

Social attributes are also essential because dealers rely on their information trading partners to provide quality insights that enable dealers to interpret market trends and events that shape their dealing decisions. In many cases, market information may be convoluted and difficult to understand, so reliable and trustworthy information partners are needed to exchange ideas on a reciprocal basis:

Trading information is a two-way street: if you want info, you must give info. You need to have a network because you don't trust everyone. You develop relationships with people you trust, and they're the ones you're prepared to share snippets of information with in constructing your opinion about what you think is going to happen in the market. It's a very clubby sort of industry. (Swaps Dealer, Australian Bank)

These informal exchanges occurred in pubs over lunch and after-work drinks. The importance of information flows combined with labour market considerations, has led to the relocation of banks from other national cities to Sydney. As an informant explained, one swap bank:

> ... relocated to Sydney from Adelaide because it was easier to replace labour and the experienced labour market is in Sydney. There is a depth of personnel that does not exist elsewhere in the country. From a market perspective, it's important to communicate with other bankers – this is where information exchange occurs and it would never happen over the phone. (Swaps Dealer, Australian Bank)

The social features of dealing networks also extend to relationships with client firms and account for the embeddedness of swap banks. Dealers stressed that greater competition and declining profitability have meant closer relationships have grown in importance:

> Most corporations are in Sydney or Melbourne, and it's important to be near them. Meeting face-to-face creates a friendlier atmosphere, and if you know them personally through meeting at a pub then it gives the encouragement to meet them directly. If you're not here, it makes it more difficult. (Swaps Dealer, British Bank)

In summary, the interest rate swaps industry demonstrates the centrality of local embeddedness in the production of sophisticated financial services that have a dealing-orientation. A key feature is that swaps are created in over-the-counter markets through electronic telecommunications infrastructure. This theoretically makes it possible for MNBs to deal from overseas locations in New York, London, or Tokyo. However, swap banks must be locally embedded to build interpersonal relationships with futures brokers and rival dealers to facilitate transactions. Despite their strong competitive mentality, dealers must routinely cooperate with each other to exchange pieces of relevant information used to construct individual views of market trajectories. Since these informally developed strategies underpin the formal market positions of banks as business corporations, embeddedness in dealing networks is needed to facilitate face-to-face meetings and establish credibility, reciprocity, and trustworthiness among dealing partners. Finally, social embeddedness also extends to client relationships.

Disembedding Brokers on the Sydney Futures Exchange

> Many global firms offer *global clearing*. A client might be smashed on its
> Australian dollar position, but making money on its Yen position. The client
> gets offset because he has cash credits in the Yen account. Whereas if he had
> his Australian account with us, we would call him for a margin call which he
> couldn't take out of his account with his Tokyo broker. Clients want ease of
> trading, so it's easier to go with a single global broker. You get one set of
> paperwork. *Back office functions are essential –it's not all glossy trading.* If
> all your trading is with one broker, it's easier than trying to reconcile
> statements from 10 different broking accounts. (Futures Broker, Australian
> Bank)

The importance of multinational corporate networks could not be more
pronounced than in the global reorganisation of the futures industry.
Despite severe competition worldwide, many foreign futures firms have
sought memberships on as many exchanges as possible. This is part of their
globalisation strategy to form an integrated futures business and one-stop
24-hour trading access to a multiplicity of localised futures markets
worldwide for their institutional clients. As the quote above indicates, local
embeddedness in the futures industry is important, but is attributable
primarily to providing back office global clearing services, rather than
front office functions. Local Sydney operations are needed to complete
globally integrated futures networks, and are not based on the importance
of the Australian futures market in itself.

The concept of local embeddedness in the futures industry operates at
numerous levels of interpretation. At the time this research was
undertaken, futures trading networks occurred with rivals on the floor of
the Sydney Futures Exchange (SFE), and the open outcry system by
definition necessitated close spatial proximity and embeddedness. At the
level of the *individual trader*, the centrality of social embeddedness in
formal business networks is nowhere more imperative than in the futures
industry. Patterns of trading between rival traders were largely determined
by interpersonal 'mateship' and underpinned by *whom* traders know:

> Most brokers survived on what can be described as an *'old boys network'*.
> After deregulation when the futures markets exploded, all these young guys
> came in as trainees – they drank together, played football together, and that's
> really how most of the relationships developed. Whether these guys were
> professional, intelligent, or could add any value was an irrelevant matter. It
> was simply a case that these guys all knew each other and drank in the same

pub three to four nights a week. That's how most of the business relationships developed in our market. (Futures Broker, Swiss Bank)

A central part of futures business is becoming socially involved with rival traders because it is required to fill futures orders, and traders that do not possess these skills cannot survive. Futures broking requires a different skill set that is much easier to acquire than the esoteric requirements of swaps dealing and master custody. Once traders have learned the hand signals, their trading capability is related primarily to their ability to cultivate a network of 'mates' on the floor of the SFE. This is significant because the floor trading relationships define formal inter-firm business networks. Consequently, social meetings remain important elements of the work process, and a small number of bars around the SFE facilitate this role.

Severe competition means that social networks are a strategy to differentiate services. This is compounded by the fact that all firms provide the same service because futures contracts are homogeneous and brokers have similar market information. Firms argue that a large part of their differentiation is based on the 'quality' of traders and client advisors. However, 'quality' is synonymous with the personal contact networks that individuals had developed. These social contacts are assets with market value, and futures personnel are often hired because of *whom they know*, as brokers explained:

You have to be seen to be better. That's tough when all firms do the same thing. This is where personality is important. If you have good people – not because they are more intelligent but just better people – this is important. Individuals are a saleable commodity, and if your competitors see that you have good relationships, then you are highly sought after. (Futures Broker, French Bank)

Futures business is relationship driven. A lot of information is the same around Sydney, so rapport is very important. It's not related to better execution - we're basically all the same. (Futures Broker, American Bank)

The standardised features of contracts, the limited nature of skills required to trade futures, combined with the socially embedded nature of execution networks have implications for creating very high turnover within the labour market. Traders change firms to increase their salaries. This has contributed to a situation whereby brokerage is the lowest in the world but the cost of personnel is very high. Firms have not trained personnel in-house but have poached traders from rivals, and this has led to higher salaries. Consequently, although they retain their floor memberships, several broking houses have rationalised their execution services and subcontract to external brokers and now focus only on clearing. The high costs and inefficiencies of the open outcry system led to the introduction of computer-based screen trading in 1999 (Anon 1999).

The importance of client advising has also declined to a certain extent. As discussed earlier, although relationships with swap dealers remain important, the globalisation of investment by pension funds, insurance, mutual funds and other collective investment vehicles, has been a major force in the globalisation of futures firms. Australian domestic institutions are important, but futures firms have expanded their marketing to target international clients. Institutional investors have their own research departments thus lessening the need for information and much advice. Instead, they just place orders for contracts. Furthermore, these sophisticated investors are able to negotiate worldwide trading commissions that place relationships at a firm, not personal level. Negotiated commissions mean that:

A lot of foreign futures firms are not earning profits on the SFE – they're subsidised by their offshore parents. They're just trying to gain market share. (Futures Broker, Australian Bank)

The Australian futures market in itself is insignificant to most multinational futures firms. But a presence on the SFE is useful because foreign futures firms:

... have an established client base and they focus on serving them with little emphasis on local Australian marketing. These firms service the Australian futures needs of offshore clients. They source their offshore business independently, or from within their headquarters and overseas affiliates. (Futures Broker, Australian Bank)

The embeddedness of multinational futures firms on the SFE results from the need to provide global clearing services. Futures firms are electronically linked to the SFE's back office to clear and settle trades and provide their international clients with a consolidated report of trading positions (and the financial gains or losses associated with those trades) on exchanges worldwide. This involves electronic data links from exchanges worldwide, to a centralised global or regional processing centre, that enables futures firms to update clients' trading positions and subsequently download these to each futures office. The alternative that non-global futures firms are confronted with is faxing numerous summary sheets from execution brokers around the world, and manually inputting these into local processing systems. Global clearing systems also enable futures firms to differentiate quality of service based on the quickness of reporting, and packaging and reporting the information in ways customised to client specifications.

The greatest benefit of global clearing is that it allows clients to have one bank account with a single clearing firm to implement *span margining* (Calder 1993). Span margining means that if a client's futures contracts lose value on one exchange, the money needed to account for the loss can be cancelled out by gains made on other exchanges. Without global clearing, a client would need to have multiple accounts in different currencies, and make such 'top-up' payments on a daily basis. Span margining and global clearing make it easier for institutional clients that trade worldwide to avoid this administrative hassle.

Despite its transactional orientation, the futures industry demonstrates that local embeddedness is related primarily to the corporate strategy of MNBs to provide back office clearing services, and less to the roles of front office trading and client advisory functions. While seeming contradictory, this does not suggest that embeddedness in trading is unimportant. Open outcry trading is a deeply interpersonal system of financial interaction, requiring social embeddedness in networks of rival traders. These social trading networks are important because the standardised characteristics of futures contracts means that the only scope for adding value is in the ability to execute an order quickly through friendly rivals. But the high costs of a trading team combined with declining commissions, have meant that for many futures firms it is more cost effective to outsource trading to other floor members. Likewise, the growing role of professional institutional investors as key clients has lessened the need for detailed advising services. Institutions undertake their own research, and use futures contracts principally as insurance against their longer-term investments. By holding onto their contracts for longer time periods, this reduces the need for frequent advising communication. However, these institutional investors demand global clearing services to ease the burden of clearing, settlements, and payments reconciliation procedures, thus accounting for the embeddedness of MNBs on the SFE.

'Geosyncretisation' in Master Custody

> The challenge for the global custodian is in providing a network of service centres all linked to the core information and processing centres. In other words the challenge is to become truly global... keeping all parties in all time zones and locations, synchronized with each other... So suddenly global custodians face the problem of a 27 hour working day – Sydney at 9:00 a.m. through to Los Angeles at 5:00 p.m.! (Hockley 1991).

As Hockley suggests, master custodians are confronted by *'geosyncretisation'* as a unique form of globalisation. Master custodians must be geographically and temporally synchronised to overcome the frictions of global and local. This is achieved by simultaneously interlinking global and local space through electronic corporate networks, thus making it impossible to speak of global and local as separate concepts. Instead, global and local are a single unified spatial entity.

From a front office operational perspective, it is necessary for master custodians to be locally embedded in Australia. But, in contrast to both derivatives industries, the embeddedness of master custodians is not attributable to local business networking with rivals or supplier firms. The most important inter-firm networks are formed with supplier sub-custodian banks. However, interaction in sub-custodian networks occurs at the global and not local scale, and electronically through SWIFT and a globally centralised securities data processing hub. Consequently, although master and sub-custodians may be spatially proximate in Sydney, they have no direct local linkages.

Interbank networking in master custody is a mechanism to deal with the complexities and expansion of securities markets worldwide. Securities markets are subject to domestic (local) regulations and have distinct local market characteristics, and strategic alliances are formed with domestic sub-custodians that act as local corporate surrogates. Sub-custodians occupy the commoditised end of the industry spectrum and provide basic securities processing functions into master custodial production chains, and their linkage into global networks epitomises a simple integration strategy. While there is intensive two-way communication flow, and sub-custodians are meaningfully integrated as essential components of global production networks, all decision-making authority, business control, and power resides with master custodians. The most substantial benefit of using external sub-custodians rather than a network of internal offices is the business power it grants in minimising costs while maintaining high levels of sub-custodian performance:

Master custodians can cut a sub-custodian out if its service is not up to standard, and they control the price and costs. Global custodians can control the quality of the service and threaten to go to another sub-custodian if it doesn't perform well. That's the benefit of using a sub-custodian network of other banks – it's difficult to sack your own people at another operation. (American Master Custodian)

Although sub-custodians are locally embedded securities processing specialists, they perceive themselves only by their inclusion as extensions in master custody networks:

Sub-custodians provide specialist services in Australian securities environment on behalf of foreign clients, primarily global custodians. A global custodian based in Boston or London for example, are dealing with over 80 markets, and it's unrealistic to expect that they will be experts in all of these. ... We are the Australian experts as far as our clients [master custodians] are concerned. (Australian Sub-Custodian)

Also, although sub-custodians are embedded in local securities markets, their linkages are also electronic following the dematerialisation of Australian securities markets in the 1990s. Corporate and government bond trading occurs through Austraclear and the Reserve Bank Interbank Transfer System (RITS). Through its Clearing House Electronic Subregister System (CHESS), the Australian Stock Exchange has implemented electronic settlement procedures. Stock market information feeds directly into the internal systems of sub-custodians, and following the settlement of trades, the system alters the records, transfers funds to client accounts, and sends out messages giving the details of the transactions. As a consequence of these developments, the importance of local embeddedness in Sydney has declined:

Until 1995 we had 20 people in our Sydney office that did several functions associated with the physical settlement of securities. However, with CHESS, we no longer have to walk around Sydney to exchange scrip with trading counterparties – it all occurs electronically, and so this can be done in Melbourne. There are only two of us here now, and we will be absorbed into other areas of the bank here in Sydney. (Australian Sub-Custodian)

The social embeddedness of business networks in swaps and futures contrasts to master custody, where it has no importance in their business.

Master custody professionals do not go for drinks after work with rivals, nor do clients select master custodians for 'mateship' reasons associated with individual personnel:

> Social networking has limited business value to us. We do not go to meet our rivals for lunch etc., and we only meet collectively at industry meetings twice a year. We are in a competitive cut-throat business and we never discuss information of any strategic business value. (Australian Sub-Custodian)

Furthermore, the most important sources of information are derived from within global sub-custody and intra-corporate master custody networks. However, these contact networks are global in scope, and occur primarily through telephone contact, electronic message systems including SWIFT, and internal communication systems, rather than through face-to-face meetings:

> I have weekly phone contact with personnel in offshore locations in our custody operations. In Hong Kong, our technology people and relationship managers; in London, marketing staff and relationship managers. (British Global Custodian)

> Interaction between offices worldwide is increasing over time. Clients are becoming more global, and therefore we must work more as a global organisation – this is a strategy and competitive advantage where we can service their business no matter where in the world they are investing or establishing operations. (American Master Custodian)

The lack of social dimensions of business is related to the characteristics of custodial banking. Unlike the transactional mentality of swaps and futures, securities administration is more complex, with relationships formed at the inter-firm rather than individual level. Investment managers contract custodians for a long-term period, because securities held by institutions involve international investments, and it is a major undertaking to transfer them to other custodians:

> Changing sub-custodians is not prohibitive in costs, but it is a time-consuming and messy process. You also need permission from your clients to change sub-custodians, and if the global custodian changes sub-custodians too frequently, then clients lose faith in the abilities of the global custodian to make a good selection. (Australian Sub-Custodian)

Despite the electronic nature of their networks, master custodians are locally-embedded for two reasons. First, relationships with client firms (portfolio managers) are central in contributing to the embeddedness of master custodians. Respondents stressed that:

> ... even though custodial services are provided electronically and could be centralised from New York, having a local presence is essential. Client service is what master custody is all about. (American Master Custodian)

Second, master custodians must be embedded to understand prevailing accounting and taxation frameworks, and the institutional features of Australian securities markets. This is related to the fact that Australian portfolio managers as clientele are very sophisticated in their investment strategies, as demonstrated by the usage of derivatives:

> Reporting on the effective exposure of derivatives is not a problem for us – [Our Firm's Funds Management] has been using derivatives for years. If clients want to do some weird swaps, we know how to account for them. But foreign custodians have had to learn these things because Australia has very particular accounting and reporting requirements. It's tough for foreign custodians to come in and build a system for CGT and AAS25. [Master Custodian A] approached us to do their tax reporting because their processing system in the US couldn't do it. (Australian Master Custodian)

Finally, the importance of the market for custodial labour varies in importance for local embeddness. Master and sub-custodians stressed that the industry is characterised by vertical integration and in-house expertise. There is a very specialised internal division of labour, comprising a range of clerks who undertake tasks in capitalisation changes, client services, reconciliations, settlement teams specialising in particular securities (i.e. stocks, bonds, cash, derivatives), accruals, income and dividends collection, and corporate actions. Although there is a high degree of internal specialisation, sub-custodians argued that they could recruit from stockbrokers, funds managers, and insurance. Consequently, the labour market is seen as ubiquitous and has enabled some sub-custodians to operate from Melbourne and Adelaide.

Access to a skilled labour market is also a central consideration in master custody, but their highly skilled labour force of primarily accountants and information technology specialists has prevented decentralisation beyond the CBD fringes:

We have 150 people but only 30 to 40 are client-specific. The rest are back office operational staff, who could be anywhere. Custodians are questioning whether they have to be centrally located. We have two floors of people here [Sydney CBD] and pay high office rentals. But if we relocated to [suburban] Parramatta, we would lose 80% of our workforce. If we moved to North Sydney, we would lose 20%. *So our labour force does exert a substantial locational pull – expertise is critical in this business.* (American Master Custodian)

Notwithstanding master custody's highly electronic basis and back office orientation, embeddedness is still important in this activity. But, embeddedness in this context is not as it is theorised in the financial services literature. Formal interbank production networks in master custody are organised as strategic alliances with domestically based sub-custodians that occur at the global not the local scale. These networks are not suffused with informal social relationships among custodial bankers. Embeddedness among master custodians remains important because master services are produced completely in-house, and this necessitates access to a skilled labour force of accounting and information technology professionals. These personnel understand the complexities of national accounting and taxation regimes, and the features of local securities markets. Although the electronic basis of custody makes it theoretically possible to decentralise operations to lower-cost suburban business parks, these skilled personnel are a centripetal force in preventing the relocation of master custodians beyond the fringe of the Sydney CBD. Also, the electronic basis of custodial banking hold the ever-present threat of inadvertently usurping client service with technological 'gloss and glitter'. Master custodians recognise this potential pitfall by building close client relationships through local embeddedness.

Conclusion

Global and local interrelationships are important in shaping the organisational dynamics of industries and firms. In financial services, processes of global restructuring, including deregulation and change in the roles of nation-states, the proliferation of technologies, and the emergence of a super-league of financial conglomerates, have introduced claims for the 'end of geography'. However, this chapter has demonstrated that these claims are exaggerated and poorly conceived. The case studies of swaps dealing, futures broking, and master custody reaffirm the importance of geography in patterns of financial production, and illustrate that embeddedness assumes different forms depending on the specific circumstances of providing particular financial services.

The financial geography literature argues that the local embeddedness of MNBs, in the financial districts of world cities, stems from formal interbank networks and their congruent informal social relationships among finance professionals. These integrated networks are central mechanisms for accessing the necessary information flows underpinning the production of financial services. This is nowhere more critical than in the swaps industry. Swaps are customised value-added transactions involving significant levels of interbank communication and negotiation based on firm-specific information. The associated dealing process involves the exchange and assimilation of localised information flows within dense networks of interlinked rival swap banks and broking firms. Trustworthiness, reciprocity, and the social embeddedness of networks assume significance because formal dealing decisions are based on the reliability of information obtained from these networks.

In contrast, the futures and master custody industries demonstrate that embeddedness cannot be explained in terms of the operation of local interbank networks. Futures contracts are characterised by homogeneous information, and consequently the ability of traders to differentiate their services is very limited. This is because of the very nature of futures business. Whereas swaps involve dealing and creating securities, futures broking involves exchanging contracts only. The social embeddedness of formal trading networks is clearly illustrated by the futures industry, where patterns of trading among rivals are largely determined by personal 'mateship'. However, the relative importance of front office trading services among some MNBs has declined in favour of back office global clearing services. A local presence on the SFE is related to the corporate strategy of MNBs to form a globally integrated network of operations to clear the global trades of their institutional investors. In master custody, interbank networks with their sub-custodians are orchestrated at the global and not the local scale, and all custodial banking services are produced in-house. The social embeddedness of interbank networks in swaps and futures does not exist in custodial banking. The lack of social dimensions of business is related to the nature of these services. Unlike the trading mentality of swaps and futures, the provision of securities services is more complex and relationships are formed at the institutional not the personal level.

Local embeddedness is also essential to foster close and trusting relationships with clients. This is important in the swaps industry because clients actively participate in the swaps creation process by providing firm-specific information that necessitates face-to-face meetings with swap dealers and clients to develop rapport and trustworthiness. Relationship management with clients is also central in contributing to the local-embeddedness of master custodians despite its electronic and back office orientation. Taxation and the institutional conventions of local securities markets make it important for custodians to remain abreast of developments to advise clients. Also, the electronic reporting basis of master custody places an even greater emphasis on embeddedness to ensure that levels of service are surpassed, as a means of differentiation from rivals. In contrast, although relationships with client firms are important in the futures industry, the homogeneous features of contracts, combined with the growing prominence of international institutions as clients, lessens the need for protracted communication and lengthy negotiations. Networks with clients are important, but many are formed at the global and not the local scale, reflecting the globalisation of the futures industry. Finally, in all three case studies, local labour markets for skilled financial personnel are an important consideration in partially accounting for local embeddedness.

This chapter has shown the dangers of making the assumption that all financial services are moving towards a seamless global network. For some, globalisation is increasing rapidly. For others, more complex embedded economies shape overall patterns, effectively resisting globalisation. A central conclusion from this chapter is that geographers need to adopt more critical analyses in examining the globalisation and embeddedness dichotomy in different financial service industries. While these three case studies have provided important insights into the universality of global-local interrelationships, further empirical research into other financial services is needed.

References

Abolafia, M.Y. and Biggart, N.W. (1991), 'Competition and markets: an institutional perspective', in A. Etzioni and P. Lawrence (eds), *Socio-Economics Toward a New Synthesis*, Armonk & M.E. Sharpe Inc., London, pp. 211-231.

Agnes, P.L. (2000), 'The "end of geography" in financial services? Local embeddedness and territorialization in the interest rate swaps industry', *Economic Geography*, vol. 76, pp. 347-366.

Aharoni, Y. (1993), 'Globalization of professional business services', in Y. Aharoni (ed), *Coalitions and Competition: The Globalization of Professional Business Services*, Routledge, London and New York, pp. 1-19.

Amin, A. and Thrift, N. (1993), 'Globalization, institutional thickness and local prospects', *Revue d'Economie Regionale et Urbaine*, vol. 3, pp. 405-427.

Anon, (1999), 'Floorless transition', *SFE Bulletin*, September/October.

Bertrand, O. and Noyelle, T (1988), *Human Resources and Corporate Strategy Technological Change in Banks and Insurance Companies: France, Germany, Japan, Sweden, United States*, OECD, Paris.

Calder, S. (1993), 'The derivative wizardry of Oz', *Futures and Options World*, July, pp. 33-37.

Casserley, D. (1991), *Facing up to the Risks: How Financial Institutions Can Survive and Prosper*, Harper Business, USA.

Clark, G.L. and O'Connor, K. (1997), 'The Informational Content of Financial Products and the Spatial Structure of the Global Finance Industry', in K.R. Cox (ed), *Spaces of Globalization: Reasserting the Power of the Local*, Guilford Press, New York and London, pp. 89-114.

Corbridge, S., Thrift, N. and Martin, R. (eds) (1994), *Money, Power and Space*, Blackwell, Oxford.

Daniels, P.W. (1986), 'Foreign banks and metropolitan development: A comparison of London and New York', *Tijdschrift voor Economische en Sociale Geografie*, vol. 77, no. 4, pp. 269-287.

Daniels, P.W. (1993), *Service Industries in the World Economy*, Blackwell, Oxford.

Daugaard, T. and Valentine, T. (1995), *Financial Risk Management: A Practical Approach to Derivatives*, Harper Educational Publishers, Sydney.

Gad, G. (1991), 'Toronto's financial district', *Canadian Geographer*, vol. 35, no. 2, pp. 203-207.

Grabher, G. (1993), 'Rediscovering the social in the economics of interfirm relations', in Grabher, G. (ed), *The Embedded Firm: On the Socioeconomics of Industrial Networks*, Routledge, London, pp. 1-31.

Hockley, J. (1991), 'Users demand more effective services', *Asian Finance*, vol. 17, no. 5, pp. 29-30.

Kindleberger, C.P. (1983), 'International banks as leaders or followers of international business', *Journal of Banking and Finance*, vol. 7, pp. 583-595.

Leyshon, A. and Thrift, N. (1992), 'Liberalisation and consolidation: The single European Market and the remaking of European financial capital', *Environment and Planning A*, vol. 24, pp. 49-81.

Llewellyn, D. (1992), 'Bank capital: The strategic issue of the 1990s', *Banking World*, vol. 10, no. 1, pp. 20-25.

Markusen, A. (1996), 'Sticky places in slippery space: A typology of industrial districts', *Economic Geography*, vol. 72, pp. 293-313.

Moss, M.L. and Brion, J.G. (1991), 'Foreign banks, telecommunications, and the central city', in P.W. Daniels (ed), *Services and Metropolitan Development: International Perspectives*, Routledge, London, pp. 265-284.

Napoli, J.A. (1992), 'Derivative markets and competitiveness', *Economic Perspectives*, vol. 16, no. 4, pp. 13-24.

O'Brien, R. (1992), *Global Financial Integration: The End of Geography*, Royal Institute of International Affairs, London.

OECD (1993), 'Financial conglomerates', *OECD Financial Market Trends*, vol. 56, pp. 13-20.

O'hUallachain, B. (1994), 'Foreign banking in the American urban system of financial organization', *Economic Geography*, vol. 70, pp. 206-228.

Peet, J. (1992), 'Financial centres', *The Economist*, vol. 323, no. 7765, pp. s3-s26.

Pryke, M. and Lee, R. (1995), 'Place your bets: Towards an understanding of globalisation, socio-financial engineering and competition within a financial centre', *Urban Studies*, vol. 32, pp. 329-344.

Storper, M. (1997), 'Territories, flows, and hierarchies in the global economy', in K.R. Cox (ed), *Spaces of Globalization: Reasserting the Power of the Local*, Guildford Press, New York and London, pp. 19-44.

Storper, M. and Scott, A.J. (1990), 'Work organisation and local labour markets in an era of flexible production', *International Labour Review*, vol. 129, pp. 573-591.

Ter Hart, H.W. and Piersma, J. (1990), 'Direct representation in international financial markets: the case of foreign banks in Amsterdam', *Tijdschrift voor Economische en Sociale Geografie*, vol. 81, no. 2, pp. 82-92.

Thrift, N. (1987), 'The fixers: The urban geography of international commercial capital', in J. Henderson and M. Castells (eds), *Global Restructuring and Territorial Development*, Sage Publications, London, pp. 203-233.

Thrift, N. (1994), 'On the social and cultural determinants of international financial markets: The case of the City of London', in S. Corbridge, N. Thrift and R. Martin (eds), *Money, Power and Space*, Blackwell, Oxford, pp. 327-355.

Thrift, N. and Leyshon, A. (1988), '"The Gambling Propensity": Banks, developing country debt exposures and the new international financial system', *Geoforum*, vol. 19, no. 1, pp. 55-69.

Thrift, N. and Leyshon, A. (1994), 'A phantom state? The de-traditionalization of money, the international financial system and international financial centres', *Political Geography*, vol. 13, pp. 299-327.

Tordoir, P.P. (1994), 'Transactions of professional business services and spatial systems', *Tijdschrift voor Economische en Sociale Geografie*, vol. 85, no. 4, pp. 322-332.

Tornqvist, G. (1983), 'Creativity and the renewal of regional life', in A. Buttimer (ed), *Creativity and Context*, Lund Studies in Geography Ser. B, Human Geography, No. 50, Department of Geography, Royal University of Lund, Lund, pp. 91-112.

Chapter 9

Local Embeddeness and Service Firms: Evidence from Southern England

Paul Search and Michael Taylor

Introduction

This study examines the extent to which producer service firms are embedded within the local economies of provincial centres in the UK. There is a growing literature that suggests that being locally embedded in business relations built on trust, loyalty and reciprocity creates the social capital that allows individual businesses and business clusters to expand and succeed through 'learning' and innovation (Maskell *et al.* 1998, Maskell and Malmberg 1999). Most of the research that underpins these ideas is based on empirical analyses of the manufacturing sector, especially of the high tech industries of Silicon Valley (Saxenian 1994, Scott 1988), the engineering industries of Baden Württemberg (Cooke and Morgan 1994, 1998), and the craft-based industries of the Third Italy (Pyke and Sengenberger 1990, Bianchi 1998). However, the growth of producer services is equally characteristic of the emerging flexible economy, with its processes of specialisation, clustering and globalisation that are producing a global mosaic of regional production centres. Producer services are supplied to businesses and governments rather than to final consumers – they are B2B rather than B2C. They are diverse and include advertising, accounting, management consulting, legal services, financial services and insurance. Their competition is said to depend on proximity to clients, competitors and other specialists, government departments and agencies, and to a diversified labour market (Daniels 1993). These services are rapidly internationalising and are coming to dominate employment in the urban economies of the developed world (Daniels 1999). There is a need, therefore, to begin to explore the roles of networks and embeddedness in

151

the operations of producer service firms to begin to understand the role of socially constructed relationships in their commercial activities.

The aim of this chapter is to begin the process of unpacking the question of producer service embeddedness by analysing the form and nature of the buyer-supplier relationships of small firm and corporate sector producer service firms in Chelmsford and Portsmouth. The first city, Chelmsford, is to the north of London, with good access to the City. Since the 1970s, it has experienced significant service sector job growth, attributable in part to the activities of the government's Location of Offices Bureau that sponsored head office decentralisation from London between the mid 1960s and the late 1970s. Portsmouth is an industrial and defence related city on the Hampshire coast. Defence manufacturing, especially in the electronics field, is important in this city, but it has experienced significant pressure through the 1980s and 1990s. That pressure has been the result of changing government attitudes towards defence procurement, the 'peace dividend' following the end of the Cold War, and the global restructuring pressures that have affected corporate sector enterprises (Mason *et al.* 1991). These processes have significantly reshaped the city's industrial and commercial environment.

The study is based on 18 in-depth interviews conducted between December 1997 and June 1998 with managers and partners at eight office establishments in Chelmsford and ten in Portsmouth. The firms interviewed were selected randomly from databases for each city developed from a range of sources. These included *Yellow Pages* and *Thomson* business directories, the *Essex Business and Industrial Directory* produced by Essex County Council, and the *Hampshire Business Directory* produced by Hampshire County Council. The interviews were semi-structured, and sought to discuss a predetermined range of issues connected with firms' commercial operations. Each interview covered four distinct areas of questioning: the firm's internal organisation; the formation of inter-firm contacts and transactions; the nature of inter-firm relationships, and exploration of the meaning of embedded relationships[1]. Interviews lasted for up to one and a half hours, though many were shorter. They were taped and transcribed, verbatim, so that the 'voice' of the respondents could be retained within the analysis.

The discussion of embedded inter-firm relationships is developed in two main sections, the first exploring transactions and relationships with suppliers and the second exploring transactions and relationships with clients. Because of the very strong differences in the nature of the inter-firm relationships developed by small firms compared with corporate sector branches, these two types of relationship are explored separately in each of the two main sections of the chapter. The mechanisms used by small firms to establish supplier relationships are explored in greater detail, owing to the depth of embeddedness involved. The study points to a fundamental division between the embedded business relationships small producer service firms experience and the arm's-length relationships of corporate sector branch offices. It is suggested that as producer service firms grow and internationalise, the managerialism they bring to local and regional economies has the potential significantly to undermine local business formation processes.

Transactions with Suppliers

Among the surveyed producer service firms, there were three distinctive types of input transaction:

- *contractor services* – building and building maintenance, electrical repairs, furniture repairs, cleaning and gardening;
- *external suppliers* – stationery, printing, office equipment, computer hardware and software, vending machines, photocopiers; and
- *professional services* – barristers, tax consultants (including VAT), insurance specialists, company formation specialists, accountants, surveyors.

Contractor and external supplier transactions are far more frequent than transactions with other professional service providers. Most frequently, in the corporate sector, professional services are provided in-house through the head office. Small firms in contrast buy in these professional services on an occasional basis and, from the evidence of the interviews, avoid them if at all possible. However, irrespective of the type of supplier transaction, the interviews demonstrate consistently different patterns of transactions for SME producer service firms compared with their corporate branch office counterparts. On the one hand, SME producer service firms have stronger, locally embedded relationships with suppliers, while, on the other hand, corporate branch offices tend to have market driven supplier relationships, most frequently mediated through and controlled by their head offices.

SME Producer Service Firms' Supplier Transactions

Small and medium sized producer service firms have classically embedded local relationships with suppliers and contractors. These relationships are, for the most part, long-standing and non-contractual, and are founded on trust, loyalty and quality. Two Chelmsford respondents summed up the situation:

> We like long-term relationships with our suppliers, on the whole. We are usually prepared to pay a little more ... service quality is very important. (firm C5)

> Most relationships with suppliers are long-term. It is because we are generally pleased with the competitiveness of the price and the quality of service. (firm C6)

Indeed, eleven of the surveyed small firms 'stick with the people we know' (firm P1), and in a number of cases they have 'stuck with them' for 25 years or more. Through these long-standing relationships, firms get after-sales service (firm P1), and special, flexible treatment from printers, for example (firm C5). These relationships save time, which was a consideration for one Portsmouth solicitor who was concerned to maximise the time available for fee-earning work (firm P4). In this case, the respondent was willing to compromise on price, but this was not generally the case. The respondent found it onerous to shop around for suppliers because, 'it's the quality of service on the product as well as price that is important' (firm C4).

Long-standing relationships go beyond trust and loyalty, they also involve shared expectations and values. The respondent at a Portsmouth accountancy firm expressed these further dimensions of embedded transactions particularly clearly:

> Where a lot of companies will write to us and offer company formation services for £70-80, we have a supplier or agent who has dealt with us for 20 or more years which charges me £189. But, I know for my money everything will go smoothly ... although it's a cost I am passing on to a client. I can guarantee there'll be no problems. However, for £70-80 they may be fine, but then again they might not be. (firm P10)

For the most part, no contracts are involved in the supplier relationships of SME producer service firms, except for maintaining computer hardware, software support, photocopiers and, in one case, the maintenance of vending machines within the office. At the heart of these relationships, 'it is trust, it is honesty. If you have got a good relationship with somebody and they have done you a good job you do not quibble about the price' (firm P4). Clearly, networking and dealing with people on a personal, professional basis to build trusting and long-term relationships is what the managers and senior partners of small and medium sized producer service firms ultimately aim to achieve. Burchell and Wilkinson (1996) have identified these relationships elsewhere in British industry. They interpret informal trust-based relationships between trading partners as involving the observation and honouring of informal, unwritten standards and understandings, within which promise keeping is central. It is suggested that these forms of relationship go *beyond* contracts. Mutual obligation transcends the contract (Burchell and Wilkinson 1996, Malecki and Tootle 1997).

Nevertheless, a consistent theme running through the interview responses is that small firms keep their supplier relationships constantly under review, and that price is central to that review process. Reflecting on their long-term relationships with suppliers, one Chelmsford firm commented that 'Loyalty is obviously a factor, but price is the most important thing. Price, followed by service, then loyalty after that. We don't tend to have trouble with suppliers so we stick with what we have got' (firm C6). The respondent from a chartered accounting firm elaborated the mechanisms of surveillance and review. 'We have had our old suppliers for years. If we think there's someone else offering a better deal then we will go and get an offer from those people and put it in front of the old supplier and say "can you match it?"'(firm C7). Again, in the words of a chartered surveyor, 'I am always aware of their price. I am always checking prices. But, I tend to be happy with them [their current supplier] and they don't try anything on. ... Also, from experience, I know the chap that does repairs and [he] is good in terms of quality and price. However, give it a few years and I might look around' (firm C3). A Portsmouth firm complained that supplier relationships do not last for more than 'a couple of years ... [because] ... they [suppliers] become complacent and we slip from being their alleged favoured price position to paying the same as everyone else' (firm P9). This particular firm broke with one supplier after it was taken over by a previous supplier who had provided poor quality service – 'We must have good quality service' (firm P9).

To some extent, these attitudes reiterate the evidence collected in the US by Granovetter (1985). In that work he found that it required a shock to jolt a buyer out of placing repeat orders with a favoured supplier. There was a cost attached to finding new suppliers and establishing new relationships. There was also less risk in dealing with established suppliers, and buyers may have established valued relationships with suppliers' representatives. In short, firms become embedded in their long-term relationships with external suppliers. The UK evidence suggests, however, that trust and loyalty is conditional. It persists only when issues of price and quality are acceptable otherwise these criteria come immediately to the forefront.

Mechanisms Used by Small Firms to Form Supplier Relationships

The mechanisms that create and mould the supplier relationships of small firm providers of producer services are quite distinctive, and demonstrate the conditional relationships of embeddedness and social capital formation in this sector. Respondents to the survey showed that two significant mechanisms operate:

* recommendation and third party referral; and
* client preference.

These mechanisms interact and are mutually reinforcing, building and strengthening the embedded business networks of a place.

Third party referrals arise when acquaintances and business partners recommend other people and companies to undertake work for a firm. This mechanism of supplier selection was virtually universal among the SMEs surveyed for this study. Thus when respondents were asked how they select suppliers they commented that: 'It is usually by word-of-mouth. Somebody knows somebody, a recommendation' (firm C7); 'Its personal experience, word-of-mouth, things like that' (firm C6); 'Recommendation is our first method [of selecting suppliers]' (firm C5); 'Ask someone who has done it before' (firm P6); and 'The answer is ... networking, networking, networking. Always go to people you know. Ask them if they know somebody they can recommend. Always do that' (firm P4).

Trust and familiarity through recommendation reduces risk and promotes stable, embedded, long-term relationships with external suppliers. As Uzzi (1997) explained in the context of the New York garment industry:

> The 'go between' performs two functions: He or she rolls over expectations of behaviour from the existing embedded relationship to the newly matched firms and 'calls on' the reciprocity owed him or her by one exchange partner and transfers it to the other. In essence the go-between transfers the expectations and opportunities of an existing embedded social structure to a

newly formed one, furnishing a basis for trust and subsequent commitments to be offered and discharged. As exchange is reciprocated, trust forms. (p. 48)

As he also explained, in a separate report on the same research:

a lack of prior social relations leaves ... [a] ... new tie without initial resources and behavioral expectations that reduce outcome uncertainty. Consequently, actors are relatively unlikely to invest, a priori, in cultivating an embedded relationship with unknown actors. (Uzzi 1996, p. 680)

But, as a businessman in Portsmouth pointed out, third party referral also serves to safeguard the continuing relationship between the first two parties. If the new relationship bears fruit that first relationship is enhanced. If it fails, it remains relatively unaffected because there is someone else to blame.

Small producer service firms also seek suppliers among their clients to build business, reinforce relationships and, in a sense, to show solidarity. As a Portsmouth accountant explained, 'We try where possible to source from clients, but not to the extent that it would compromise our professional relationship' (firm P1). Put more bluntly by a Chelmsford accountant about his firm's relationships with a stationer, 'They are local, we're customers of theirs and they're clients of ours' (firm C6). Firms use clients as suppliers where appropriate, '... because its beneficial for them as well as us' (firm P2). 'We are very much reliant on local business to support us [as clients] and where we can we try and reciprocate' (firm P10). To convert a building for their own use twelve years ago, a Portsmouth accountancy firm depended entirely on its clients. 'We acted as our own main contractor and used all clients to build and refurbish [the offices]' (firm P3). The architect, builders and a team of sub-contractors (electricians, plumbers and painters) were all local clients of the firm.

Through these mechanisms long-term relationships of trust and loyalty are created and further reinforced even if their origins become lost in the mists of time. 'We have been here for 25 years and we tend to use the same firm again and again. How we made the original contact I just don't know. I guess that most people that work for us are clients of the firm. I know the electrical repair man is a client, the stationery supplier is a client. So, that gives us a broad base of suppliers, to use our own clientele' (firm C8).

The embeddedness of small business service firms goes further than using clients as suppliers and building links through third party referral. Firms in both Chelmsford and Portsmouth see themselves as part of a local business community. In effect, they are aware of being part of a place-specific 'milieu' (Maillat 1995, Crevoisier 1999) which goes beyond the individual firm. Respondents expressed this awareness in a number of ways.

> [Portsmouth] is a strong professional community both in a social and private sense. (firm P1)

> There is a strong professional community in this area [Chelmsford] and we are very well known in that. (firm C5)

> We are a local firm with local connections and we enhance those, and an ultimate aim is to enhance the community as well. (firm C4)

> It's a lot of camaraderie. You are all there for the same purpose – helping each other in the business community. (firm C6)

Being part of a local business community is not always a positive experience. Though generally positive, that local community can also be intrusive. A Portsmouth accountant complained that the local professional community is too closely intertwined (firm P1), and a firm of solicitors (firm P4) broke with a local accountant because they did not want the rest of the local business community knowing their business.

Corporate Sector Branches' Supplier Transactions

The nature of the transactions developed by corporate sector branches with contractor services, external suppliers and other professionals stand in stark contrast to the close, locally embedded relationships of small firms. Five corporate sector branch offices were covered in the study, and the nature of their transactions for these types of input was principally dependent on the levels of autonomy the branches enjoyed within the larger organisations to which they belonged.

In the four least autonomous branches (firms C1, C2, P5 and P7), most supplies were organised by the head office through formal contracts with other national companies. Contracts with approved suppliers covered stationery, computer hardware and software, photocopiers, furniture, vending machines, catering, recruitment and training, and, in one case, building work and maintenance (firm C1). Professional services were provided from panels of experts or they were provided by the head offices themselves; financial control, risk control, business and legal services, for example. As one branch manager put it, 'Any things that may affect the company as a whole are dealt with at our head office' (firm P5). Only over minor building work and maintenance and cleaning was there any local control, and for a Chelmsford finance firm, '[t]he only other major firms we deal with are recruitment agencies' (firm C1), and only then because that function was being devolved to the branch which was being 'grown' within the larger organisation. But, even when these purchasing functions were delegated to the branch level, they were unlikely to embed the branch within its local community because, in at least one case, using a local supplier involved presenting estimates to a regional manager for approval (firm P5).

In the least autonomous branches there was clearly a level of frustration on the part of some managers about not being able to manage. 'I have delegated power ... [but] ... I can't run the business as I would perhaps think best' (firm C1). The respondent at a Chelmsford firm remarked that the power to nominate local service suppliers had been taken away from branches on a lot of occasions because of the perceived bulk buying power of the head office (firm C2). The same respondent pointed out that this was not always the case. Negotiations they had concluded with a local cleaning firm who was already known to them proved to be 40% cheaper than the price negotiated by the head office with another supplier.

In contrast with this picture of branch subordination, one Portsmouth branch office enjoyed considerable autonomy (firm P8). The operation had been set up in Portsmouth in the late 1980s by a large insurance company to undertake loss adjustment work for insurers, brokers and corporate clients that self-insure. At the beginning of 1998, in a bid to cut costs, the operation was sold to a large firm of loss adjusters. When the manager was interviewed, the operation functioned largely as a semi-autonomous profit centre within the new firm. The interview showed how a shift in management style, involving increased autonomy, can begin to embed corporate providers of producer services in local communities, mirroring significant aspects of small firms in their relationships with suppliers. Central to selecting suppliers was 'historic and personal experience' and third party referral. 'I think it is quite important to use recommendation. It always helps if someone has experienced something before and they pass it on to me' (firm P8). Some links with contractors and external suppliers were long-term, but also conditional. 'We are reasonably loyal to our present existing suppliers, but I do tend to invite other people to send me their [information]. Price is extremely important as too is quality'. These are hard-headed decisions devoid of the altruism that Uzzi (1997), for example, has identified in embedded inter-firm relationships:

> You have got to be able to build strong relationships and partnerships with suppliers, which is what we like to do ideally, because you are in a better

negotiating position and [have] a good understanding. ... Over a longer period of time the custom we give our suppliers increases, which gives the opportunity for bigger discounts. (firm P8)

The inescapable conclusion from this small sample of branch offices of corporate sector producer service providers is that most do not engage with the local business communities within which they are located. For contractor services and external supplies they are under strong central control. When that control is relaxed, however, some elements of local embeddedness appear to emerge.

Transactions with Clients

For transactions with clients, similar forms of relationships are again apparent for small producer service firms compared with corporate branch offices. While the former are long-standing and strongly embedded, the latter are market driven and controlled centrally within each corporation. The differences are very strong, and hold significant implications for the growth and sustainability of business in provincial towns in the UK, especially as corporate sector service firms expand along with their market-driven relationships.

SMEs and their Client Relationships

Every small firm surveyed as part of this study described their client base as being essentially 'local', meaning for the most part within a 30-40 mile radius of the firm. 'Local', however, was perceived in different terms in different firms. One Chelmsford firm saw 'local' as falling within a two-hour drive time (firm C5), while a Portsmouth firm (firm P3) described it as covering Hampshire and West Sussex. For others 'local' was far more constrained, just Portsmouth in one case (firm P9) or the Portsmouth-Chichester area in another (firm P4), which is less than 20 miles. For two firms, that local area contained 80-95% of their clients (firms C6 and P10), and for two others it contained 75% of their clients (firms C8 and P6). Notwithstanding the geographically constrained nature of most firms' client bases, many respondents still spoke of having significant numbers of nationally-based or internationally-based clients.

Relationships with clients are also almost always long-term, making them close and special. When asked if they had long-term relationships with their clients, a partner in a Chelmsford firm of solicitors (firm C5) explained:

> Yes, very much so. [A] very high percentage of people we have instructed before – 70-80%. ... Sometimes you find that as your clients' firms grow their empires expand and, if you relocate out of the area, they still stay with you because they are happy with the service. ... It is very important that your clients like you, and you like them. You can't carry on for very long if there is no sort of chemistry between people.

The same respondent went on to explain how strong social relationships between firms can be mutually beneficial by expanding the business between them:

> ... we like to explore business friendships. Certainly, the lead partner of a company can be a good friend of ours in business terms. ... [O]ther people in his organisation may seek advice from other departments [in our organisation] – property, wills, employment litigation. So, you get a lot of connections between his firm and ours, lots of individual strands, and it is important that they are all kept in good repair. It builds a strong relationship. (firm C5)

Other survey respondents expanded on these themes. A Portsmouth respondent reinforced the importance of personal chemistry in shaping client relationships because of its role in building trust:

> ... because 90% of my clients only understand only 40% of what I do for them. If they don't trust me, two things happen. One, they are on my case, which is a pain in the behind. And two, they run the costs up because they are always phoning up asking what is going on – please explain. (firm P3)

The respondent in another Portsmouth firm expanded on the links that develop between businesses as they develop, indicating strong symbiotic relationships between firms in local business communities:

> We tend to get quite involved with ...[our clients] ... and more so than other firms of accountants would tend to do. All our big clients were start-ups 10 years ago. So, we have a very hands-on role with financial directorships, consultancy, and helping them buy and sell businesses. (firm P3)

He went on to explain that the firm's clients are principally family-owned businesses and that, '... all our big clients have been with us since the start of the business', a point reinforced by an unrelated Portsmouth accountant (firm P2).

Underpinning these long-term, symbiotic relationships between strongly locally embedded firms is the issue of goodwill – established custom – which might be appropriately described, in this particular context, as 'the personal chemistry of knowing'.

> [P]eople's knowledge of ourselves and our history is very important. That is the goodwill aspect. We deal with people's accounts, their audit, their tax. And knowledge of what happened five to seven years ago is actually quite important. (firm P10)

At the core of goodwill is loyalty, and there was loyalty in abundance evident in the surveyed small firms' relationships with clients. Indeed, when a Chelmsford accountancy firm closed a London office in 1996 in response to economic circumstances, they expected to lose some clients but '... retained 70-75% of the work they were handling in London. They [clients] followed our partners who are now based in Chelmsford' (firm C6).

These characteristics of small firms' client relationships closely replicate those found in New York by Uzzi (1997), being '...distinguished by the personal nature of the business relationship and their effect on economic process' (p. 41-42). They imply multiple ties between firms, with the potential for any one dyadic relationship to run both ways, with clients as suppliers and suppliers as clients. 'The ties of each firm, as well as the ties of the ties, generate a network of organisations that becomes a repository of the accumulated benefits of embedded exchanges. Thus the level of embeddedness in a network increases with the density of the embedded ties' (Uzzi 1997, p. 48). In Grabher's (1993) terms, these small producer service firms are not only structurally embedded in their local business communities, but cognitively and culturally embedded too. Not only are they tied locally as consumers and producers, they all work on the same information base and share common values and attitudes to doing business. For the most part, the small producer service firms surveyed for this study are inextricably linked, both socially and commercially, with their local business community.

Corporate Sector Branches and Their Client Relationships

In comparison with their small firm counterparts, the links that corporate sector branch offices develop with their clients are far more straightforward and far less subtle. In this respect, they again reflect the nature of their supplier transactions, putting them at arm's-length from their local communities.

Of the corporate sector branches surveyed as part of this study, two were wholly concerned with providing other businesses with their particular types of service. The first, an office in Chelmsford (firm C1), is involved in securities administration, as the custodian of securities and as the trustee of authorised unit trusts. It is a classic 'back office' that administers the business of 40 commercial clients but does not generate business itself. That task is performed elsewhere in the larger organisation. As such, the branch is not engaged in creating strong client relationships of any type, locally embedded or not. Its ties into the Chelmsford community are minimal, through a limited set of input transactions and through the local people it employs. The second office, in Portsmouth (firm P8), belongs to an international firm of insurance loss adjusters. Its client base is spread across the UK and abroad and comprises:

> [S]everal hundred insurers, large firms of brokers and small firms of brokers and certain corporate clients as well who may be self-insured and who want to use our expertise to either bulk handle small claims or build up large losses that they may not have recovered, large fines or serious motor accidents.

Though their client base is large, they claim they try to build strong, long-term relationships with their clients. The manager pointed out, however, that while new communication and information technology means they are free to locate anywhere and still build those strong relationships, the practicalities of leases and their review dates tie the operation to specific places for specific blocks of time. He went on to link these pressures to his professional service needs and the adequacy of the Portsmouth business environment for his type of firm:

> Our product is the management of insurance claims. ... The types of insurance claims we handle tend to be quite specialist in nature, often involving litigation. We have close relationships with solicitors as well, so we have to be a little careful about where we locate because we need the relevant staff and technical experience. ... And in some places staff are just not available. Portsmouth happens to not be a particularly good place to be (firm P8).

These views stand in stark contrast to carefully nurtured local client relationships revealed among small firm, business service providers in both Portsmouth and Chelmsford. Although it may, in part, reflect the limited specialist expertise in the UK's provincial cities, it also reflects the wider geographical orientation of corporate sector managers and their limited local engagement.

The remaining corporate sector offices (C2, P5 and P7) surveyed as part of this study are primarily engaged in selling standardised insurance products to people in their homes – motor insurance, household insurance and travel insurance. They are in effect regional offices selling standard products to a local clientele in their homes or over the phone. Contact with clients through their office was, in fact, positively discouraged by one firm (P5). Selling is mechanical and governed by company guidelines. Under these circumstances, the office can be located anywhere. There are no locally embedded client relationships. They are footloose and organisationally subordinate, with one Portsmouth respondent suggesting that their operations might conceivably be relocated to the outskirts of Southampton.

The evidence presented here, though slight, points to the branches of large producer service firms having few, if any, embedded client relationships when they are sited in provincial cities. They are in those locations as a result of historical accident, to cut costs on office rentals, to escape paying employees a London allowance, and to be able to sell to clients in their own homes. They are not in those locations to feed off and feed a local business community.

Conclusion

The analysis developed in this chapter demonstrates the radically different nature of the transaction structures of small firms compared with corporate branch offices engaged in the provision of producer services. Among the surveyed small firms, a firm's clients can be its suppliers and vice versa. Supplies and services are bought from local firms and contractors. In turn, those local firms buy professional services. Professional work is also passed within the local community of professionals. At the same time, these transactions are not arm's-length, they are close-in and personal. They are long-term and involve trust, loyalty, reputation and reciprocity. They are also conditional. Though they are not price-determined, they are still price-sensitive. They are also governed by quality. In short, the buyer-supplier relationships of small firm providers of producer services are classically locally embedded. Significantly, here is the local embeddedness that Curran and Blackburn (1994) suggested, in a major empirical study, had died among SME manufacturers in the UK.

But, embedded relationships go further than the orchestration and manipulation of transactions to satisfy the social obligations and rules of governance in a local business community. Because of the very nature of the professional services small firm, accountants, surveyors and solicitors provide, and because of the assets they control in trust, they play a key role in processes of local new firm formation. In those processes they may provide advice or they may equally participate as active investors or as investment angels. In this respect, small producer service firms are not just the glue of a local business community, they are also to some extent the drivers and facilitators of that community, with an important role in promoting local prosperity. This, however, is not the strength of weak ties that Granovetter (1985) saw in the shifting transaction structures of industrial clusters and that is now seen as the foundation of local learning and local innovation (Maskell *et al.* 1998, Asheim 2000). This is the *strength of strong ties*, which are as strong as cross-investment and risk sharing in a local community. When small firms are interpreted as temporary coalitions of individuals seeking to exploit opportunities for personal wealth creation (Taylor 1999, 2000), small firm, producer service providers play a key role in coalition formation. The evidence presented here is, however, limited. Nevertheless, it suggests hypotheses worthy of further exploration.

Among the branch offices of corporate producer service firms, in contrast, locally embedded relationships of any sort are few and far between. By and large, they are company mediated market ties or internally provided services. In these organisations, market mechanisms and the internal command economy have replaced symbiotic, embedded relationships. It can be conjectured that such relationships are too complex and difficult to maintain in large organisations as they strive to achieve corporate goals and coordinate the activities of large workforces playing on national and international game boards. The rules and guidelines of managerialism have replaced the symbiotic embeddedness of small firms. Indeed, following Johannisson (2000), it can be suggested that the advance of managerialism, that inevitably follows concentration in the producer service sectors, has the potential to seriously erode local processes of new firm formation.

166 Embedded Enterprise and Social Capital

Note

1   The '*internal organisation*' section of the interview sought information on how the firm was organised, the numbers and types of service providers and external suppliers it used and the age of the business. The section on '*forming inter-firm contacts*' examined how contacts are made, the use of written contracts and the propensity of firms to source locally or serve local clients. Questioning on '*inter-firm interaction*' was concerned with such issues as price, quality and service, long-term relationships and how good performance is rewarded. The '*embeddedness of relationships*' was examined through the questions of the strength of the tie between the firm and the locality, the frequency of use of local services and the issue of 'community'.

References

Asheim, B. (2000), 'Industrial districts: The contributions of Marshall and Beyond', in G. Clark, M. Feldman and M. Gertler (eds), *The Oxford Handbook of Economic Geography*, Oxford University Press, Oxford, pp. 413-431.
Bianchi, G. (1998), 'Requiem for the third Italy? Rise and fall of a too successful concept', *Entrepreneurship and Regional Development*, vol. 10, pp. 93-116.
Burchell, B. and Wilkinson, F. (1996), *Trust, Business Relationships and the Contractual Environment*, WP 35, Economic & Social Research Council, University of Cambridge.
Cooke, P. and Morgan, K. (1994), 'Growth regions under duress: renewal strategies in Baden Wurttemberg and Emilia Romagna', in A. Amin, and N. Thrift, (eds), *Globalization, Institutions and Regional Development in Europe*, Oxford University Press, Oxford, pp. 91-117.
Cooke, P. and Morgan, K. (1998), *The Associational Economy*, Oxford University Press, Oxford.
Crevoisier, O. (1999), 'Innovation and the city', in E. Malecki, and P. Oinas, (eds), *Making Connections: Technological Learning and Regional Economic Change*, Ashgate, Aldershot, pp. 61-77.
Curran, S. and Blackburn, R. (1994), *Small Firms and Local Economic Networks: The Death of the Local economy?* Paul Chapman Publishing, London.
Daniels, P.W. (1993), *Service Industries in the World Economy*, Blackwell, Oxford & Cambridge, MA.
Daniels, P. (1999), 'The overseas investment of US service enterprises', in D. Slater and P. Taylor (eds), *The American Century: Consensus and Coercion in the Projection of American Power*, Blackwell, Oxford, pp. 67-83.
Grabher, G. (1993), *The Embedded Firm. On the Socioeconomics of Industrial Networks*, Routledge, London.
Granovetter, M. (1985), 'Economic action and social structure: the problem of embeddedness', *American Journal of Sociology*, vol. 91, no. 3, pp. 481-510.
Johannisson, B. (2000), 'Modernising the industrial district: rejuvenation or managerial colonisation?', in E. Vatne and M. Taylor (eds), *The Networked Firm in a Global World: Small Firms in New Environments*, Ashgate, Aldershot, pp. 283-308.
Maillat, D. (1995), 'The territorial dynamic, innovative milieus and regional policy', *Entrepreneurship and Regional Development*, vol. 7, pp. 157-165.

Malecki, E. and Tootle, D. (1997), 'Networks of small manufacturers in the USA: Creating embeddedness', in M. Taylor and S. Conti (eds), *Interdependent and Uneven Development: Global-Local Perspectives*, Ashgate, Aldershot, pp. 195-221.

Maskell, P., Eskelinen, H., Hannibalsson, I., Malmberg, A. and Vatne, E. (1998), *Competitiveness, Localised Learning and Regional Development*, Routledge, London.

Maskell, P. and Malmberg, A. (1999), 'Localised learning and industrial competitiveness', *Cambridge Journal of Economics*, vol. 23, pp. 167-190.

Mason, C., Pinch, S. and Witt, S. (1991), 'Industrial change in southern England: A case study of the electronics and electrical engineering industry in the Southampton city-region', *Environment and Planning A*, vol. 23, pp. 677-703.

Pyke, F. and Sengenberger, W. (1990), 'Introduction', in F. Pyke, G. Becattini, and W. Sengenberger (eds), *Industrial Districts and Inter-Firm Co-operation in Italy*, International Institute of Labour Studies, Geneva, pp. 1-9.

Saxenian, A. (1994), *Regional Advantage: Culture and Competition in Silicon Valley and Route 128*, Harvard University Press, Cambridge MA.

Scott, A. (1988), *New Industrial Spaces: Flexible Production Organization and Regional Development in North America and Western Europe*, Pion, London.

Taylor, M. (1999), 'The small firm as a temporary coalition', *Entrepreneurship and Regional Development*, vol. 11, pp. 1-19.

Taylor, M. (2000), 'The firm as a temporary coalition', Paper presented at the workshop on Conceptualising the Firm in Economic Geography, University of Portsmouth, 11 March.

Uzzi, B. (1996) 'The sources and consequences of embeddedness for the economic performance of organizations: the network effect', *American Sociological Review*, vol. 61, pp. 674-698.

Uzzi, B. (1997) 'Social Structure and Competition in Interfirm Networks: The Paradox of Embeddedness', *Administrative Science Quarterly*, vol. 42, pp. 35-67.

Chapter 10

Embedded Project-Production in Magazine Publishing: A Case of Self-Exploitation?

Carol Ekinsmyth

Introduction

Recent perspectives on local economic development, focusing on notions of 'embeddedness', 'clustering' and 'learning regions' give priority to inclusionary and symbiotic relationships between economic actors in networked systems of production. Often these conceptualisations assume, implicitly if not explicitly, a bounded locality in which fairly enacted and mutually favourable relationships between economic enterprises create localised, positive externalities. Underpinning these inclusionary collaborative relations are trust, reciprocity and loyalty between economic actors, that work to create social capital and competitive advantage both for firms and for the regions in which they operate (Granovetter and Swedberg 1992, Braczyk *et al.* 1998, Asheim 2000, Porter 1998).

The argument developed in this chapter is that this set of propositions needs to be seriously questioned and qualified in the context of the labour market conditions associated with the local embedding of enterprise. Two aspects of the model are questioned and criticised in the argument developed here. First, the over-emphasis placed on the benevolent aspects of economic exchange is questioned, together with the attendant failure of theory adequately to incorporate unequal power relations between interacting parties, especially in the relationship between business enterprises and the labour they employ. Second, the reluctance of the model to look beyond the spatial boundaries of 'the firm' to fully recognise a wide range of social and labour market problems associated with spatial clustering.

The empirical focus of this chapter is magazine publishing, one of the increasingly important cultural industries in Britain. The industry has gone through a period of significant change over the past twenty years, with increasing corporate concentration of ownership (McKay 2000, Stokes and Reading 1999, Gall 1998), union de-recognition and significant organisational streamlining. The consequence has been an increasing externalisation of risk, costs and sunk costs, involving the shedding of employees and increasing reliance on subcontractors and freelances. Indeed, the industry can now be interpreted as operating a form of 'organisation as project' (Ekinsmyth 2002, Grabher 2001).

Seemingly strong processes of social and spatial embedding, in the form of networks of social relations, underpin 'project' forms of organisation (Grabher 2001). However, it is argued here that the embeddedness model with its emphasis on the benevolent aspects of social relations has limited use in this context. Instead, project members (journalists, editors and production teams) operate within a corporate architecture that trades on worker self-exploitation. Far from mutually beneficial relationships involving trust, reciprocity and loyalty, it is the unequal power relations between corporations and their labour that creates competitive advantage. In short, there is a more sinister side to embedded relationships, that is under-researched in the literature yet highly visible in magazine publishing. It results from the corporate cooption of the social processes vital to cultural industries. It can be suggested on the evidence presented here that the embeddedness model of local growth has failed to recognise this element of cooption because it has failed to fully engage with issues of unequal power, both between firms and between firms and the labour they use.

Flexible production in magazine publishing has at its core the externalisation of costs. These costs can be divided into those that are recoverable and those that are not (sunk costs – see Clark and Wrigley 1995). In the former category are recoverable employment overheads or office space costs. In the latter, are two types of sunk costs that are particularly significant: employment sunk costs, in the form of superannuation and redundancy benefits; and capital sunk costs, in the form of office equipment and furnishings, building maintenance and alterations. Increasingly, these costs are externalised as much of the productive activity in the magazine industry occurs in the London homes of freelances. The pushing of office functions into residential areas puts new pressures on urban areas. It means the crowding of residential space and the inflation of local property prices as office demand joins residential demand to drive the system. In this way, there is a blighting of residential life as well as a shift in the urban real estate market. At the same time, the boundaries between work and home are blurred for workers and their families, with the attendant stresses this creates. So far, the embeddedness literature has failed to look beyond the boundaries of 'the firm' into the private lives of workers and their families, to understand the wider implications of locational clustering for the local areas in which clustering takes place. Beyond notions of local wealth creation, local competitive advantage and 'local social capital' (Putnam 1993), there has been little investigation into what is occurring in the localities, homes and lives of those people who are supposedly benefiting from enterprise embeddedness and clustering.

The information on which this chapter is based was collected through open-ended interviews with 41 magazine industry workers, including 19 freelance workers, 20 in-house staff (who are mostly Sub-editors, Editors or Publishers) and two trade union officials. Interviews were mostly carried out in 1999 and contacts were made by gradually infiltrating the social networks of individuals and snowballing from there. The sample includes individuals working in all major areas of endeavour in magazine production, editorial, publishing/budget management, personnel, freelance sub-editors (who generally work in-house on day rates), freelance writers (who generally work from home), freelance PR and trainers (many freelances were multi-skilled as part of their business strategy). So far, the project has not extended to include advertising staff or to move up through organisations to board-level. All interviewees worked mostly for London-based projects, even though two freelances lived in the north of England. All others lived centrally within London.

The industry is young in terms of the age profile of its workers. Whilst the average age of interviewees was 30, actual ages varied from 22 to 51. The sample tends to be biased towards more experienced workers. Interviewees worked in various magazine sectors, ranging from health journalism to computer games magazines, home and lifestyle magazines, women's glossies, music magazines, men's glossies. Some freelances worked for newspapers and television, as there is some, though limited, crossover between the two sectors.

The argument developed in this chapter is divided into five sections and is based largely on the information gathered through interviews[1]. Following the introduction, Section 2 outlines the organisational characteristics of the magazine industry, drawing on notions of the 'project' form of business organisation. Section 3 shows how project forms of organisation in magazine publishing rely on the self-exploitation of a labour force, including both its employed and freelance components. Section 4 considers the spatial dimensions of project organisation in magazine publishing, particularly the implications of the externalisation of sunk costs. Section 5 takes the exploitation argument further by showing how spatial clustering in the industry has a darker side. Indeed, the empirical example discussed here suggests that the embeddedness literature has seriously underestimated the importance of both unequal power relations and space, as factors shaping the nature and orientation of economic exchange.

Magazine Publishing

British magazine publishing is a dynamic, rapidly evolving industry, heavily concentrated in terms of ownership and geographical location. Through merger and acquisition, the sector has seen an increasing concentration of ownership by large multinational and transnational corporations, often with interests wider than the media industries alone. At the other end of the spectrum are a number of smaller, national companies that have enjoyed success and growth in recent years[2]. Alongside corporate change, over the past 10 to 15 years there has been significant change in the organisation of production and associated working practice. In common with other cultural industries, magazine publishing has a fluctuating, unstable and unpredictable market, added to which, the market has seen a dramatic increase in size in recent years. Indeed, commentators estimate there to be around 10,000 titles produced in Britain today (Stokes and Reading 1999). Companies have responded by making their operations flexible, and externalising sunk costs, most notably by shedding permanent employees and making heavy use of freelance labour. In this way, individual issues of magazines are produced by temporary coalitions of actors, some of whom are employees and some of whom are freelance, in an operational framework that is better conceived of as a 'project' than as a conventional production unit within a firm.

The magazine industry is not alone in adopting this form of organisation. In a discussion of the structure of London's advertising industry, Grabher (2001) demonstrates that project forms of organisation dominate, and discusses why they have a competitive edge. He argues that in particular it is their 'soft assembly' (p. 358) that enables a creative, flexible and responsive approach to business, where, '… higher levels [of the organisation] subsume the activity of lower organisational layers by controlling it only in a limited way. Such "soft assembly" allows lower organisational units to respond to local contexts and to exploit intrinsic dynamics' (Grabher 2001, p. 358). The assembly of a team of people with a mix of skills appropriate to the specific job in hand enables an exact fit between people and purpose, and the temporary nature of working relationships enables a sharing of experience, ideas and practice that engenders creativity. This, according to Grabher, gives Soho advertising agencies a competitive edge over their more traditionally organised competitors. It requires a locality endowed with a latent network of skilled workers who can be brought into projects as and when they are needed. Grabher suggests in his conclusion that this form of project-based organisation is becoming more widespread than advertising and the cultural industries alone: '... advertising is practising and refining forms of project organisation, whose paradigmatic importance goes far beyond the culture industries' (Grabher 2001, p. 371).

Project-organisation in magazine publishing has similarities with, and differences from, the Soho advertising agencies, and much of this is due to the nature of the product. There are arguably more continuities between projects in magazine publishing than there are between advertising campaigns, thus perhaps requiring a slightly less variable team composition. In the magazine industry, each single issue of a magazine can be considered as a project. There is variation between titles rather than between companies in the extent to which the project form of organisation is used. Weekly titles tend to have a larger in-house staff and make more repeated use of particular freelances. Monthlies and titles produced less often, can exercise more flexibility in their project teams, and typically have a higher ratio of freelance to in-house workers. In common with advertising, the CEOs and upper levels of companies for the most part play only a minor role in the day-to-day control or functioning of project teams. They devolve this responsibility to the project teams and an individual, often called a Publisher, who has control of budgets for one or more titles. For each magazine edition, the Publisher will decide the targets for advertising revenue and title sales, agree the ratio of advertising to editorial space for the issue, instruct the advertising team on their targets, and inform the magazine Editor about the editorial space to be filled. Advertising teams are not normally dedicated to specific titles, but like the Publisher, often work on a number of titles. Magazine Editors, however, are dedicated to one title, and it is they who have overall responsibility for the editorial material, graphics and the final putting together (or the 'putting to bed', as it is known in the trade) of an issue. The Publisher is an in-house worker. Advertising teams are also normally employees. Editors too are normally employees, but for some smaller titles, the Editor can be freelance.

Project and editorial teams in magazine publishing normally consist of an Editor, an Editorial Assistant, Sub-Editors for various editorial categories (for example, fashion, health, features in women's magazines), and an Art Director responsible for graphics. This team is often employed, but in some cases, the whole team can be made up of freelances. For each edition of a magazine, the editorial team decides on the content of the edition, and then produces or commissions that content. Most of the top consumer titles do very little in-house writing but, instead, commission writing from the freelance writing community. Similarly, photography is normally commissioned, and picture research may also be commissioned. Freelancing sub-editors will normally work in the magazine's offices. Writers and photographers work at home. Sub-editors gather and edit all copy and graphics, and typesetting is carried out. Once all the material has been gathered and approved, the issue is sent to outside agencies for printing and distribution.

Whilst the situation might be different in advertising, in the magazine industry this form of organisation certainly has far-reaching consequences for workers for whom corporate flexibility often equates with personal insecurity. Editorial teams work to tight budgets in fickle markets. The business is hard-nosed. Costs must be kept to a minimum. Titles can fail very rapidly and workers, staff and freelance, are kept on a knife-edge of employment insecurity. Such an environment breeds worker compliance, even where this acts against the interests of the worker community. Over the past twenty years, the magazine industry has changed dramatically. New technology, de-unionisation, the shedding of employees in favour of freelances, massively expanded but increasingly volatile markets, have all acted to create an environment in which all workers are over-worked, experience high levels of employment-related stress, and are not in many cases entirely happy. Many respond to this situation by becoming freelance. Others are propelled into freelancing by redundancy. Once freelance, most discover that the stress and over-work are different but still present.

Projects forms of organisation are temporary coalitions of people that have a limited lifespan and a limited purpose. In magazine publishing they consist of all the contributors (editorial, advertising, marketing, publishing, printing and distribution), to a single edition of a magazine. Many project members have no contractual relationship with the magazines that they work for, and are hired by sub-editors through the casual mechanisms of social contacts and networks. Social networks and ties thus constitute the core mechanisms through which projects are constructed and deconstructed, and within this system, power, trust, reciprocity and friendship alliances are vital components and conduits of activity. The magazine industry and its workers are deeply embedded in latent, and importantly, localised inter-personal networks.

Whilst the spatial consequences of project-based organisation in magazine publishing are discussed elsewhere in this chapter, it is important to emphasise here that magazine publishing in the UK is highly concentrated in London and, for the most part, in Central London[3] (Ekinsmyth 1999, Driver and Gillespie 1993). This high degree of spatial clustering is shared with the country's other cultural industries (Pratt 1997, Driver and Gillespie 1993). The reasons for this are simple but multi-faceted. Perhaps most fundamentally, the cultural industries feed off up-to-the-minute information, news, fashions and cultural events. They benefit from being located in major world metropolitan areas (Storper 1997) that, for the most part, provide the stage for these events. Highly dynamic and stimulating environments act as a magnet to leading-edge economic activities, which tend to form tight clusters in these areas. Around them forms a pool of workers with the human capital necessary for these industries. In the case of magazine publishing, a workforce with a detailed knowledge of, and ability to read and interpret, cultural signs and signals is especially vital. The magazine industry needs to be located within that workforce. As Scott (1997) details, there is a recursive relationship between place, culture and economic activity. Whilst such spatial concentration is beneficial to capital and the local economy, Section 5 of this chapter shows that, for workers, there is a darker side to spatial clustering in the magazine industry.

Embedded Self-Exploitation

Social networks are a foundational component of the embeddedness thesis, and within these, positive and inclusionary social relations involving trust, reciprocity and mutual support are generally supposed to exist. Evidence from the magazine industry, however, reveals a less inclusionary and less mutually beneficial set of relationships associated with project-based forms of organisation, but relationships that still operate through social and friendship networks. Indeed, it suggests that the embeddedness thesis overplays the positive aspects of social relations between economic parties, by affording too little attention to the role and exercise of unequal power in shaping these social relations (Taylor 2000, Granovetter 1985). It is argued that new forms of relationship between capital and labour, of the sort that are evident in project forms of organisation in magazine publishing, constitute a new form of self-exploitation in labour markets. Specifically, embeddedness manifests itself, not as an inclusionary and mutually beneficial way of doing business, but as an exploitative form of social relations between capital and labour (see Ekinsmyth 2002 for a fuller discussion of the conceptual bases of these terms).

Self-exploitation has different dimensions for employed workers and freelances in magazine publishing, with the most obvious forms occurring among freelances. Self-exploitation is encouraged by the informality with which workers are engaged or commissioned, which relies on networks of social contacts, often friendship networks. Although there is some formal advertising and appointment of individuals to vacant staff positions on magazines, the more normal route for new appointments is through informal channels of contacts. Thus, friends and acquaintances appoint each other, and the scope for formal, disengaged discussion about appointment terms and conditions is reduced, possibly to the disadvantage of the appointee. Similarly for young hopefuls, as entry into the industry depends upon being known and who you know, it is necessary to work for little or nothing at first to get a foot in the door (Ekinsmyth 2002). Even more obvious is the disadvantage that informal friendship networks impose upon freelances. All interviewees remarked upon how vital social networks were for their work:

> ... in the whole media industry, it's all built around teams and socialisation ... networking is very important. Not even like aggressive, soliciting-type networking, but just like being seen, being out there. You know, management spend a fortune out of their own pocket buying people drinks, you know. (T, Employed Group Publisher)

> You do realise very quickly that, as a freelance, you have to be very proactive. I mean, freelance work is about who you know and relying on being known, because people – you know – if someone is commissioning and they know you are a known quantity, you don't have all those hurdles to cross first. (A, Freelance Editor)

> Every single job I've had, whether freelance or not, has come through people I know. It's such a network thing, it's a huge network thing. (S, Freelance Sub-Editor and Writer)

There is considerable scope for exploitation within a system revolving so totally around social networks. First, the costs of network construction and maintenance are almost entirely borne by the worker, not by capital (quite literally in the case of the manager speaking above). It takes time, effort, emotional energy and money to keep relationships buoyant, comfortable and effective. For freelances, this effort is vital. Second, business conducted under the mantle of friendship can result in reluctance on the part of the less powerful party (normally the freelance) to drive a hard bargain. Through embarrassment and reluctance to put the friendship and thus future work at risk, freelances often find it difficult to negotiate good rates of pay for their work. The quote below is typical of freelance feelings about rate negotiation, and demonstrates that most will accept whatever is offered rather than negotiate for more:

> I prefer not to negotiate rates – it's best if there is a company set rate. (L, Freelance Sub-Editor)

One interviewee highlighted the freelance's lack of power in this relationship:

> ... the worst organisation I have dealt with is [name of organisation] where the fee just goes up and down, and it seems on the whim of the Editor that week. And I think what they have is a budget for the week and then they share it out depending how many contributors there are, and they never tell you in advance so you take the work on spec, and then find you've got 'two-and-six' [12.5 pence – pre-decimal currency, signifying small almost worthless amount] at the end. (A, Freelance Editor)

Indeed, unequal power relations between capital and labour are underpinned by this social way of operating. Project-based organisation requires a large pool of flexible, available and affordable labour, and project members who are able to assemble these teams rapidly. It also relies on the financial and personal insecurity of freelances, where it is in the interest of capital to prevent individual freelances becoming too confident of their own worth, and thus becoming difficult or uncooperative. By placing emphasis on personal characteristics and likeability, working cultures revolving around sociability are able to erode or undermine the true economic value of individual freelances, rewarding them instead for their cooperation and dependability rather than their, perhaps more valuable, skill and ability. Although a feature of the magazine publishing sector as a whole, this aspect of self-exploitation appears, perhaps not surprisingly, to be more prevalent in consumer magazines than in the more serious news-type titles. Thus, in interviews, consumer magazine freelances emphasise their likeability more than their skill. One interviewee, speaking about in-house freelancing said:

> ...then there is the personal side of, 'do you fit in'? Erm, are you quiet enough, loud enough, friendly enough, obliging enough, forceful enough or modern enough? How do you dress? What does your hair look like? (S, Freelance Graphic Designer)

Being nice, trying to 'fit in', trying to conduct a business relationship whilst adhering to the conventions of friendship, all place additional burdens on the freelance and can be considered as a form of emotional labour. It keeps freelances unsure of their own value and abilities and, indeed, interviews with Publishers revealed that to some extent, this is a deliberate strategy on the part of those responsible for budgets on magazines. Publishers that were interviewed talked in terms of the 'upper hand', the 'balance of power' between freelance and magazine, and the dangers for the magazine of freelances becoming too powerful or vital to the commercial success of the title. It seems likely that informal friendship channels keep freelances in their place.

However, the mantle of friendship is shallow. Not far from the surface, distrust rather than trust characterises inter-personal relationships. Workers in the industry recognise that it is a hard-edged business, governed by tight budgets, profit margins, competitors and the whims of consumers. In this insecure working environment, individuals do not really trust one another and, certainly, power relations are recognised. Freelances especially know that their relationships with commissioning editors and the magazines they work for are only as good as their last piece of work, and because of this, they are under pressure to always return their best work. Equally, they are under pressure to be flexible and amenable. Many freelances complained of over-work. Many found it impossible to obtain the right amount of work, and often, this meant taking on too much for fear of losing the goodwill of commissioning editors. They were commonly approached at the last minute to write a piece or cover employee absences in-house, and most felt unable to disappoint the pleading Editor at the other end of the phone. In this way, social relationships once more result in freelance self-exploitation. Speaking of why he would never consider freelancing, one Publisher said:

> ... what a life: in and out of buildings; at the mercy of the team, being asked to work late. You get paid for everything – great – but the other thing is, there is always this pressure never to go on holiday, and I know so many freelances who have taken one holiday in three years. And it's mostly – it's not because they can't afford it – it's very often because the magazines are so desperate for them to do the work that they feel that it will injure their relationship with that magazine if they let them down. (T, Employed Group Publisher)

Similarly, freelancers spoke of over-work;

> A client will ring you up and ask you to do a job that takes three quarters of the week. Then somebody will offer you a job that takes half the week, and because you need the money, you take on both, which means you work over the weekend. (A, Freelance Writer)

While the discussion in this section has explored those aspects of worker self-exploitation most likely to result from socially embedded project organisation, Ekinsmyth (1999) has detailed many other aspects of unequal power relationships between freelances and the companies they work for. Under the theoretical mantle of Beck's (1992) *Risk Society*, she concluded that, in addition to the factors discussed above, the combination of uncertain work and pay, lack of employment benefits such as sickness, maternity or holiday pay, difficulties of taking on financial commitments such as mortgages or private pension plans, and the difficulties of working at home, all lead to an insecurity amongst freelances that acts to the advantage of employers. Ekinsmyth (2002) also shows that in-house employees in the industry are equally insecure. Indeed, by making their operations flexible and by moving to project-based forms of organisation, the magazine industry has effectively transferred many of the risks of enterprise to its workers. And, alongside risk comes insecurity (Ekinsmyth 1999, Beck 1992).

In sum, therefore, whilst at first glance the reliance of the industry on social networks might appear to offer an ideal empirical context with which to support the tenets of the embeddedness thesis, we have begun to see in this section that an underestimation of the importance of power inequalities in social relations between parties can lead to an understanding of embeddedness that is too optimistic. Whilst we can see that trust and reciprocity do play a part in the magazine industry's dealing, very often it is mistrust, power play, exploitation and self-exploitation that determines the rules of the game. As we shall see in the final sections of this chapter, there is also a geographical dimension to this unequal relationship.

Externalised Costs and Spatial Clustering

Spatial clustering need not be a necessary condition of project-based organisation in the magazine industry. New communication technologies have revolutionised the production of magazines and made it possible, in theory at least, for project members to work from anywhere in the world. In practice, however, both capital and labour demonstrate a high degree of spatial clustering, and this geographic concentration, necessary to sustain the culture of sociability, also has its exploitative aspects.

The embeddedness literature gives little consideration to the negative aspects of spatial clustering, preferring to view regional clusters as growth poles in which firms do business with each other in a mutually beneficial way. Implicit in this view is the assumption that negative externalities, caused by spatial concentration, are out-weighed by positive economic factors that benefit the region and its population as a whole. This is the perspective of local economic development officers and successful firms, but when the circumstances of workers are considered, in magazine publishing a different picture emerges. As we have seen above, by making their operation flexible, magazine companies have transferred the risk of enterprise to their workers. As part of this process, they have externalised costs where possible and this includes the costs of office space and capital sunk costs for office equipment, furnishings and so on. By transferring the location of much of the physical activity of magazine production to a freelance labour force that mainly works from home, firms save money in office space and the overheads involved in maintaining workspaces. But, relying on what amounts to a 'just-in-time' system of project formation, in which the goodwill and late availability of freelances is used frequently to bail the pared-down in-house system out of trouble (standing in for sick employees in an in-house system with no slack capacity), firms depend on the security of a pool of spatially proximate freelances. Whilst in theory, homeworking freelances could live anywhere (and many interviewees suggested this), in practice, subtle culturally embedded practices ensure that the majority stay in, or close to, the Capital. One such practice is the importance placed upon socialisation, face-to-face contact and friendship. As one Editor stated:

> ... it's very convenient if you commission something at the last minute to get it back very quickly via email, but you don't get to develop a personal relationship with the freelance, which you sort of need in order to get the most out of both of you ... I would deplore a world where I only got copy delivered by email and I never saw the whites of freelances' eyes. (D, Employed Editor)

Another interviewee refuted the 'death-of-distance' thesis:

> No, I've written many articles about this phenomenon, death of distance and all that, and yeah, I'd say it doesn't happen. I mean a lot of it still depends on being able to have stuff couriered round, or to have a beer with an Editor, that sort of thing. (M, Freelance Journalist)

It seems that while firms have been keen to externalise sunk costs, they still desire to maintain control over *where* productive activity takes place. Instead of relinquishing power, firms extend their power into the residential areas, homes and personal lives of their workforce. As the final section of this chapter demonstrates, in the magazine industry at least, clustering and spatial embeddedness have an exploitative dimension.

Loss of Residential Amenity and the Home as Workplace

The movement of office functions into residential areas is a trend that is difficult to quantify, but commentators agree that it is becoming increasingly common (Henley Centre 1998, Jackson and Van Der Wielen 1998). As many teleworkers are unable to move to larger homes when they start to work from home, homeworking often leads to the crowding of residential space. At the city scale, this pushing of work into homes can lead to a blighting of residential space, reducing the amenity of already crowded urban residential areas. This trend might also be expected to lead to inflation in house prices, as it is office plus residential demand that is driving the system. This is especially problematic for workers with less secure incomes (such as freelances) who already find it difficult to commit to regular mortgage payments (Walker 2000, Wilcox and Ford 1997, Maclennan 1994).

Homeworking freelance interviewees were quick to list a number of disadvantages of working at home, and space restrictions were high on this agenda. Many interviews were carried out in freelances' homes, typically in inner London areas such as Hackney, Clapham and Stoke Newington. As interviewees ranged in age from 22 to 51, the small flats and houses that they lived in were not a function of early adulthood. Equally, many interviewees had partners who worked, so theirs was not the sole household income. Despite this, high London house prices meant that they could only afford to live in places that were too small for the comfortable co-existence of paid work and home life:

> Space is a problem... We would like to move but our freelance lives aren't secure enough at the moment to get a mortgage. It's kind of ironic because if the two of us worked in an office, this flat would be perfectly big enough ... But when you're working from home, you've got less money but you need more space. (W, Freelance Writer)

One freelance in his fifties lived in a flat comprising one room, kitchen and bathroom. He and his partner slept in a double bunk bed with office space beneath. A settee and television sat in another corner of the room. Another interviewee worked at her computer in a lobby between the kitchen and the back door to the garden. Every time she sat back in her chair, she disturbed the cat food bowls that stood against the wall behind her. She spoke of her dream to leave London:

> I would love to leave London and have more space, but I would feel that it is more difficult to stay at the centre of things if you're not physically located here. It's just that people think, 'they're not in London, they're somehow peripheral'. (B, Freelance Writer)

Interviews revealed that space restrictions render work more intrusive in family life. If work cannot be physically separated from living space, it is more difficult for the worker to switch off from that work at the end of the day. One respondent spoke of the sinking feeling she has each time she lays eyes on her piles of papers. She likened it to that 'Monday morning feeling' that constantly overwhelmed her throughout her leisure time. Indeed, discussion in the literature has been divided on the different kinds of stresses present in home-based waged work (Moore 1999). Evidence from space-poor London freelances suggests that the continual presence of work in the home can lead to work-related stress and even, as reported by some interviewees, 'workaholism'. Family life too, can suffer in this situation.

The dimensions of home-based paid work are multi-faceted and have only been touched upon in the above discussion. It is clear, however, that the networks of social ties that underpin production in the magazine sector rely on spatial concentration and as such, are geographically embedded. Whilst certain types of freelance journalist are free to live wherever they choose, the majority feel it necessary to remain close to the spatial heart of the industry, and they do this, despite considerable disadvantage to themselves and often their families. Project-based organisation in the industry, and the externalisation of capital sunk costs relies on the continued self-exploitation of a group of workers in this way.

Conclusions

Discussion in this chapter suggests that the embeddedness literature has so far underplayed the significance of the less benevolent aspects of economic exchange within embedded systems of production. This has arisen because of a failure in this literature to fully appreciate the consequences of unequal power relationships in the social systems on which embeddedness and clustering are built. Social and spatial embeddedness in magazine publishing operates through fundamental power inequalities between capital and labour. This chapter has shown that the key mechanism in the operation of unequal power is the culture of sociability. It is this culture, and the networks of social contacts that result, that determine the rules of the game. Key amongst the 'rules' of this culture are the importance of 'friendship', and the supposed qualities of friendship such as mutual trust, support, and reciprocity – all qualities of embedded social systems. However, as the discussion in this chapter has shown, these 'rules' are under the control of the most powerful parties in social relationships (in this case capital in the form of commissioning editors), but not the control of those in the less powerful positions (freelances who have little bargaining or persuasive power). Instead of trust and reciprocity, in these embedded relationships, there is mistrust and self-exploitation.

Project-based organisation in magazine publishing appears to require the spatial proximity of its project members, despite new technologies that could enable more literal 'distance-working'. Instead, there is reluctance on the part of the industry to lessen the importance placed on face-to-face contact. This works to the favour of capital as it reinforces the importance of social networks and the power-inequalities discussed above. It ensures that, whilst some control of production is relinquished by using more home-based workers, control is nevertheless extended into the homes of freelances in the local area. Whether true or not, the perception is that project-based organisation is more manageable if its participants are spatially clustered. This has potentially detrimental consequences for freelances, their families and the residential areas in which they are located.

The magazine industry demonstrates high degrees of spatial and social embedding, but not of the type, or for the reasons, heralded by the embeddedness literature. Instead, a more sinister side to embeddedness has emerged from the discussion in this chapter. It is one that revolves around the exploitative power bases of the social networks and relations upon which this particular industry relies, and whose reach extends beyond the boundaries of the firm, out into the homes, families and residential areas of its workers.

Notes

1   Quotes from interviews cited in this chapter are followed in brackets by an anonymised initial and the job function of the interviewee.
2   The magazine sector can be divided into three: consumer titles, business-to-business (or trade) titles and contract publishing. In the UK consumer market, McKay (2000) lists the major players as IPC, EMAP, National Magazine Company, Conde Nast, D.C. Thompson, H. Bauer, G&J, Readers Digest, BBC Worldwide and Future (in order of size/importance). For the business-to-business sector, Reed Business Information, Emap Business Communications, Haymarket Business Publications, Miller Freeman and VNU Business Publications are the leaders (McKay 2000, Consterdine, 1997). Redwood Publishing and Forward Publishing are important in the contract sector, where companies tend to be smaller and nationally-based, but who produce titles that are significant in terms of circulation and readership (e.g. Redwood Publishing produce AA Magazine, Sky TV Guide and A Taste of Safeway - all in the top 5 consumer/customer titles by circulation in 1997 (Audit Bureau of Circulation Jan-June 1997, referenced in Stokes and Reading, 1999).
3   There are exceptions. Titles that are less dependent on latest fashions or trends can be in other UK locations: for example, some divisions of EMAP are in Peterborough, and D.C. Thompson is in Scotland (see Driver and Gillespie 1993).

References

Asheim, B. T. (2000), 'Industrial Districts: The contributions of Marshall and beyond', in G.L. Clark, M.P. Feldman and M.S. Gertler (eds), *The Oxford Handbook of Economic Geography*, Oxford University Press, Oxford, pp. 413-431.

Beck, U. (1992), *Risk Society: Towards a New Modernity*, Sage, London.

Braczyk, H.J., Cooke, P. and Heidenreich, M. (eds) (1998), *Regional Innovation Systems. The Role of Governance in a Globalized World*, UCL Press. London.

Clark, G. and Wrigley, N. (1995), 'Sunk costs: A framework for economic geography', *Transactions of the Institute of British Geographers, New Series*, vol. 20, pp. 204-223.

Consterdine, G. (1997), *How Magazine Advertising Works II*, Research Report 40, Periodical Publishers Association, London.

Driver, S. and Gillespie, A. (1993), 'Information and communication technologies and the geography of magazine print publishing', *Regional Studies*, vol. 27, pp. 53-64.

Ekinsmyth, C. (1999), 'Professional workers in a risk society', *Transactions of the Institute of British Geographers, New Series,* vol. 24, pp. 353-366.

Ekinsmyth, C. (forthcoming 2002), 'Production in projects: Economic geographies of temporary collaboration, Project-organisation, embeddedness and risk in magazine publishing', *Regional Studies*, vol. 36, no. 3.

Gall, G. (1998), 'The changing relations of production: Union derecognition in the UK magazine industry', *Industrial Relations Journal*, vol. 29, pp. 151-161.

Grabher, G. (2001), 'Ecologies of creativity: the Village, the Group, and the heterarchic organisation of the British advertising industry', *Environment and Planning A*, vol. 33, pp. 351-374.

Granovetter, M. (1985), 'Economic action and social structure: The problem of embeddedness', *American Journal of Sociology*, vol. 29, pp. 481-510.

Granovetter, M. and Swedberg, R. (eds) (1992), *The Sociology of Economic Life*, Westview Press, Boulder CO.

Henley Centre (1998), *The NeXt Generation: Lifestyles For The Future*, The Henley Centre, London.

Jackson, P. and Van Der Wielen, J. (eds) (1998), *Teleworking: International Perspectives: From Telecommuting to the Virtual Organisation*, Routledge, London.

Maclennan, D. (1994), *A Competitive UK Economy: The Challenges for Housing Policy*, Joseph Rowntree Foundation, York.

McKay, J. (2000), *The Magazines Handbook*, Routledge, London.

Moore, J. (1999), 'Tensions in home use and experience', Paper presented to the Housing Studies Association Conference, September.

Porter, M.E. (1998), *On Competition*, MacMillan, London.

Pratt, A. (1997), 'The cultural industries production system: A case study of employment change in Britain, 1984-91', *Environment and Planning A*, vol. 29, pp. 1953-1974.

Putnam, T. (1993), *Making Democracy Work: Civic Traditions in Modern Italy*, Princeton University Press, Princeton NJ.

Scott, A. (1997), 'The cultural economy of cities', *Institute of Urban and Regional Research*, vol. 2, pp. 323-339.

Stokes, J. and Reading, A. (eds) (1999), *The Media in Britain: Current Debates and Developments*, Macmillan, Basingstoke.

Storper, M. (1997) *The Regional World: Territorial Development in a Global Economy*, Guilford Press, New York.

Taylor, M. (2000), 'Enterprise, power and embeddedness: An empirical exploration', in E.

Vatne and M. Taylor (eds), *The Networked Firm in a Global World: Small Firms in New Environments*, Ashgate, Aldershot, pp. 199-234.

Walker, R. (2000) 'Insecurity and housing consumption', in E. Heery and J. Salmon (eds), *The Insecure Workforce*, Routledge, London, pp. 210-226.

Wilcox, S. and Ford, J. (1997), 'At your own risk', in S. Wilcox (ed), *Housing Review 1997/98*, Joseph Rowntree Foundation, York, pp. 34-46.

Zenith Media Worldwide (1997), *UK Media Handbook*, Zenith Media, United Kingdom.

Chapter 11

Local Embeddedness, 'Institutional Thickness' and the State Regulation of Local Labour Markets

Simon Leonard

Introduction

Essential to the developing critique of the embeddedness model has been the examination of the central proposition that local economic growth is fostered through trust and reciprocity in inter-firm relationships, thereby creating and building 'social capital'. Often left implicit in these analyses has been the belief that embedded and collaborative social relations between capital and labour are equally important, enabling capital to release the tacit knowledge that labour controls (Brusco 1996). For these relationships to generate local economic growth a third component, namely institutional support through government training, education and other agencies and organisations, is also seen as important to create and facilitate the conditions in which the potential of inter-firm and capital-labour relationships can be realised.

Comparatively little has been written about the nature and function of this institutional support as a mechanism for promoting local economic growth, that has gone very much farther than the arguments developed by Amin and Thrift (1994, 1995) under the heading of 'institutional thickness'. This chapter seeks to critique and develop these ideas on 'institutional thickness' by looking at the dynamics of institutional support in a specific empirical context, namely through the provision of state-funded skills training facilities within Greater London. Through an historically informed analysis, this chapter contributes to the developing critique of the embeddedness model, offering an alternative perspective on the role of these selected institutions of labour market regulation and governance in the creation and support of local economic growth. In their original presentation, Amin and Thrift (1994) state that institutional thickness involves a strong institutional presence of firms, financial institutions, local chambers of commerce, *training agencies*, local authorities, development agencies, trade unions, and other government and quasi public bodies. Number, variety and diversity of institutions were seen to constitute a necessary but not sufficient condition for the creation of local economic growth. In addition, three other elements of institutional support needed to be present: (1) high levels of interaction between agencies in a local area; (2) well-defined structures of domination or patterns of coalition in order to socialise costs and control rogue behaviour; and (3), a mutual awareness among these institutions that they are involved in a common enterprise (Amin and Thrift 1994, p. 14).

Amin and Thrift suggested that these three determinants of institutional thickness would, in the most favourable cases, produce a number of outcomes key to the promotion of local economic development. Amongst these outcomes were: first, *institutional persistence*, in which local institutions are reproduced; second, *institutional flexibility*, whereby organisations in a region have the ability to both learn and change; and third, a sense of *inclusiveness* that is the development over time of a widely held common project which, 'serves to mobilise the region with speed and efficiency' (p. 15). Ultimately, what was most important was not the presence, number or diversity of local institutions, but the processes of institutionalisation that were seen to underpin and stimulate entrepreneurship and consolidate the local embeddedness of industry (Amin and Thrift 1994).

Studies that have sought to identify the presence and relevance of institutional thickness have not unexpectedly identified a particular geographical region (often defined in terms of its neo-Marshallian significance) and then attempted to unpack local institutional diversity and interactions in order to specify the conditions under which local labour market and local economic institutions support local economic growth (Cooke and Morgan 1994, MacLeod 1997). This perspective on institutional involvement within local economies has generated important findings although, as empirical method, it is perhaps inconsistent with many of the relevant processes that underpin local economic development.

Most importantly, this regional and institutional snapshot is at odds with many of the dynamic outcomes of institutional thickness suggested by Amin and Thrift. Their original formulation specified a set of dynamic interactions that are not appropriately analysed through a locally-defined, essentially a-temporal and a-historic institutional cross-section. Ideas such as institutional persistence (reproduction), institutional flexibility (learning and changing), and the consolidation of a sense of inclusiveness, all suggest dynamic interactions and intersections between local labour market and local economic agencies and organisations in a manner that is best studied over time and without pre-specifying the significance of any particular spatial scale, from the 'local' to the supra-national.

An institutional cross-section, without an historical perspective and with a pre-determined spatial scale, makes a number of assumptions about how local labour market institutions function. First, these institutions are assumed to have consistent goals and objectives and to have consistent and persistent relationships with other local institutions engaged in supporting and facilitating economic development. Second, focusing attention at the local scale also presupposes that these same institutions may only be concerned to promote 'local' economic outcomes and that their institutional role remains tied to this scale over time. Third, the ability of these same local labour market and local economic institutions to change and to contribute at different times to national and supra-national aims and objectives is obscured by this cross-sectional perspective. Fundamentally, it over-simplifies the intimate, complex and changing relationship between the state and the economy (Block 1994). It freezes that relationship in time, and places it exclusively within one particular regime of accumulation, governance and regulation. In so doing, it severely limits and constrains our ability to interpret the shifting economic significance, both spatially and temporally, of labour market and economic institutions. What experience shows, however, is that the relationships between government and the labour market in shaping, creating and maintaining institutions of labour market regulation have always been strongly time specific. Local labour market dynamics are just that, dynamic.

The strands of this argument on institutional thickness are developed and exemplified in this chapter by reference to five time periods that represent distinctive periods of labour market regulation and governance in Greater London. The accounts focus upon one set of policy initiatives, aimed at the provision of adult skills training opportunities in Government Training Centres (GTCs) and later Skillcentres throughout Greater London. These examples do not offer an exhaustive record of this training initiative in London (see Leonard (1999) for a detailed account), but they do provide a clear statement of how institutions are reproduced, and yet change their purpose and role over time in different periods of regulatory need. Sometimes the same agencies are seen working to meet the needs of local business, and sometimes working against those needs to meet other institutional and governmental goals and objectives at local, regional, national and even supra-national scales. At any one time, and also at different times, these goals are a complex mix of economic, social and political policies: to deliver strategic national manpower planning policy; to combat national, regional and inner-city problems of unemployment and social unrest; to promote and support business and enterprise at national and local scales; and to fulfil ideological and political aims through deregulation, marketisation and privatisation (Leonard 2001).

Following this introduction, Section 1 shows how in the 1920s and 1930s, the GTC programme was used to meet both the needs of the depressed regions and the buoyant London economy through 'industrial transference'. Section 2 illustrates how these same GTCs, contributing to post-war reconstruction in the 1940s, had major implications for the support of business and the unemployed in the London economy of the 1980s. Section 3 examines London's Skillcentres in the 1970s and highlights the conflict in purpose between inner-city and national manpower planning policy in the UK at that time. Section 4 details a further and different era of conflict in purpose between national and 'local' institutions of labour market governance within Greater London during the 1980s; and finally, Section 5 details the closure of these same labour market institutions – marketised and eventually privatised in an attempt to more directly serve the needs of local business.

Labour Market Regulation and Governance in Greater London

The history and geography of the relationship between skills formation and the institutions of labour market regulation and governance in Greater London is one of a changing response to regulatory need over a period of more than 60 years. The account presented here has three principal threads running through it. First, it demonstrates 'institutional persistence' in the context of skills training initiatives, but persistence in the absence of consistency of purpose or role. Second, it demonstrates the heavy constraints placed on institutional flexibility and the ability of institutions to learn and change by the complex agenda shaped by economic, social and political processes operating beyond the 'local' scale. Finally, it demonstrates that the relationship between the state and the market is fundamental to understanding the circumstances through which local economic growth is encouraged and supported. Far from promoting growth and development, institutions of labour market regulation and governance are shown to have the potential to be in conflict 'locally', even working counter to the expressed local needs of business and enterprise. In these circumstances building a sense of 'inclusiveness' and mobilising other agencies and organisations around a mutually agreed project of promoting local economic growth is an extreme over-simplification of the way in which labour market institutions operate over time within any particular labour market setting.

Industrial Transference and Industrial Growth

In the first selected period of labour market regulation that affected Greater London, some of the earliest state-funded training centres were established in and around the capital between 1929 and 1931. Three training centres were established at Park Royal in West London and Carshalton and Waddon in South London. The Park Royal training centre, although located in an industrial area on what was then the edge of the London metropolitan built-up area, was primarily concerned with providing training for unemployed workers from the depressed mining areas across Britain. In choosing locations for these new training centres, the relative buoyancy of the local economy in which a particular GTC was to be located was of prime importance. Carshalton operated in support of the GTC scheme as a Transfer Instructional Centre. Under this initiative, workers were again transferred from the depressed regions[1], but they were deemed to be ill-prepared for the industrial training and the work regime offered at the early GTCs. The final GTC in the London region was located at Waddon (near Croydon) in South London. Waddon remained as a GTC and Skillcentre for over 50 years until its closure under the Manpower Services Commission's (MSC) Skillcentre rationalisation plan of 1984. In 1930, when the arrangements were being made for opening a new GTC at Waddon, '... the policy of selecting sites for new centres away from the depressed areas in parts of the country where industrial development was taking place was maintained' (Ministry of Labour 1931, p. 32).

A number of general and important points are apparent from the specifics of these examples. First, this 'landscape' of labour regulation and governance was created out of an emerging national regional policy designed to alleviate problems of surplus labour in one set of localities, namely the 'depressed mining areas'. The other side of these attempts to alleviate high unemployment in these regions was the buoyant local economy in London during the 1920s and 1930s. The decline of the heavy industries in the North and the associated decline in the country's mining communities, were part of the declining position of Britain in the world economy. The new growth industries, based on the domestic market, were in large part located close to London, particularly in outer north-west London which, during the interwar years, was the most rapidly expanding industrial sector of London (Martin 1966, Pratt 1994a). While the national picture was one of economic decline, the 'local' circumstances of the London economy facilitated the opening of GTCs, as part of the emerging regional policy, but also as a means of increasing labour supply to the buoyant industry of the Southeast.

The location of the three London GTCs, therefore, closely mirrored the growing industrial areas of London during the 1930s. These areas were primarily outer west and northwest London, the Lea Valley in outer north London and the area around the Wandle Valley in outer south and southwest London. All of the early GTCs and the TIC were located in these areas. Park Royal in the northwest, Waddon and Carshalton to the south and southwest; and just outside London early GTCs were located in Slough to the west, and Watford to the north. Pratt (1994a) details the political economy of these early industrial estates, including Park Royal and Slough. He illustrates how their growth and location was in part linked to these broader industrial and sectoral changes in the British space economy, as well as the restructuring of the organisation of production, based upon the principles of mass production and its associated space-extensive needs.

In terms of institutional support for the promotion of local economic growth, these early GTCs were already being cast as serving both economic and social roles, by 'supporting' both the buoyant economy of London and the depressed economies of the mining regions. After the end of the First World War, the continuing fear of social unrest, generated by economic depression and high unemployment, prompted government to respond through these GTC initiatives to meet national, regional and local political, economic and social objectives. Although it persisted through the 1930s, the GTC programme was recast in the 1940s to meet other labour market needs under different regulatory conditions of wartime.

Post Second World War Reconstruction

The numbers of GTCs across Britain during the Second World War varied according to the needs of the war effort. By 1946, the number of GTCs in London had increased significantly in order to meet both the post-war needs of reconstruction and resettlement. This period saw an increase to nine GTCs across London. Waddon and Hounslow, from the pre-war period, continued to provide training. Although training continued at Park Royal, the Park Royal centre was new, built in 1947 to meet the new regulatory need for skills training in the building trades to facilitate reconstruction. Of the other six GTCs, centres at Barking and Enfield were established at the end of the war, whilst the new centres at Alperton, Kidbrooke and Twickenham were open by August 1946, followed by Barking Annexe in late 1946 (Ministry of Labour 1947).

These locations were, at that time, most appropriate for meeting the building trades skills training needs associated with the reconstruction of the post-war London area. Training in the engineering trades had been effectively suspended at the end of the war, and GTC training was turned over almost exclusively to skills training in the building trades. Consistent with the new need, these centres were accessible to Central London. However, they required space-extensive sites to facilitate training, not factory-based locations. A number of the new post-war GTC facilities, however, were located in areas of rapid inter-war industrial growth, such as at Alperton in outer northwest London, Twickenham in west London, and Enfield in north London, in the Lea Valley area.

The demand for skilled workers in the building trades was considerable and nationally many of the new GTCs were only intended to meet this perceived short-term need and so were established as Emergency Training Centres (ETCs). A number of these building trade ETCs became permanent GTCs over time, so that many of the London GTCs at this time, opened to support the demand for skilled labour in the construction industry, remained as GTCs until the late 1980s. Even allowing for the closure of many of the London Skillcentres in the period up to the 1960s, the geography of this form of labour market regulation in London, based on the post-war priority of reconstruction, had established a locational pattern which would characterise Skillcentre training in London in the 1980s. This institutional persistence continued despite changing market regulation, industrial restructuring and shifting social needs.

Institutionally, the GTCs as policy programmes were persistent, but with little consistency in purpose and role. New and distinct roles were most apparent between the 1930s and 1940s. In addition, the GTCs were at this time 'flexible' in terms of their response to changing regulatory needs as the changing demands for skilled workers within the London economy were apparently being met by the changing skills provision. However, two problems were emerging. First, these changes were governed by a range of policy objectives that were intended to meet different economic, social and political goals. This diversity of purpose established a severe limit on the later role of the GTCs and Skillcentres in terms of their ability to respond to the expressed skills needs of local business and industry. It also severely limited the ability of the GTCs and Skillcentres to later contribute to an inclusive institutional response for the promotion of local economic development. Second, the post-war emphasis upon the building trades, created a skills infrastructure and environment that increasingly grew away from the skills training needs of many London-based employers. The loss of this 'flexible response' and the inability of the Skillcentres to learn and change within their economic environment was perhaps the most important source of government's later dissatisfaction with these training initiatives.

National Manpower Planning Versus the Inner-City Problem

The low point of GTC provision came in the 1960s. However, by 1977, skills training provision had grown into the national Skillcentre network, involving the re-establishment of a number of GTC facilities in London, and the opening of a new centre in the inner-city area of the East End of London. This shift followed the setting up of the Manpower Services Commission, by a Conservative government, following the Employment and Training Act 1973 (Ainley and Corney 1990, Evans 1992). This Act stated that a national training agency was needed to promote training in occupations that cut across industrial boundaries. The new MSC symbolised the need for an economy-wide coordination of skill formation. But, as Sheldrake and Vickerstaff (1987, p. 46) argued, it '... failed to retain or create mechanisms for translating these national policy objectives to the level where training actually occurred - the individual company'. Under the new Labour government of 1974, the MSC's original five-year plan also recognised, both nationally and within the major metropolitan areas, a conflict between the long-term policy objective of developing a comprehensive manpower planning strategy for Britain, and the shorter-term need to combat increasingly high levels of unemployment in particular geographical areas.

Whilst the MSC recognised the problems of developing a national manpower planning strategy in the shadow of rapidly rising unemployment, its analysis of that problem was resolutely fixed at the national level. The growth of the Skillcentre programme, towards the end of that Labour government, reflected that national economic objective and not the social and local labour market needs of inner London. Within London, the 'local' labour market's regulatory need, associated with the growing inner-city problem, was for a new configuration of Skillcentres. But, the national manpower policy framework at that time emphasised meeting the requirement for skilled workers, and not any particular 'local' problems of unemployment within London.

This emphasis was not simply an issue of placing the 'national' before the 'local'. More fundamentally it represented a misconception that the Skillcentres could symbolise and represent a national 'network' when in practice their value had always been closely linked to local labour market conditions. At the end of the 1970s, whilst the national skills training priority was economic, the local labour market situation in London meant that the London Skillcentres were increasingly being looked to as a source of skills training for disadvantaged groups within the London labour market. However, a disadvantaged inner-city working population was to be trained from sites based predominantly in outer London. This situation graphically illustrates the progressive lack of synchronisation between the training offered at Skillcentres, that was conceived and formulated at the national level, and the training needs of workers and employers where industrial change and processes of labour market segmentation were rapidly changing the structure of London's labour market.

This conflict in purpose represented yet another change in the role of the Skillcentres. They were directed by a national agenda that did not allow these training centres, in any particular local labour market context, to learn and change to support and promote local economic growth. As a labour market institution, the Skillcentre network was part of a developing national manpower planning strategy that was inflexible in terms of its ability to meet expressed needs from local employers. From this national perspective, the Skillcentres demonstrated persistence but not consistency of purpose. They were inflexible and resistant to changing needs, and they were 'dislocated' from other 'local' agencies of labour market regulation, unable to build or contribute to a sense of inclusiveness around a commonly held project of local economic development.

National and Local Institutions of Labour Market Governance

Within London, during the 1980s, while significant changes emerged in the industrial and employment structure of London, manufacturing industry continued to decline (Kowarzik *et al.* 1989, Leonard *et al.* 1991a, Leonard *et al.* 1991b, Pratt 1994b). The London Skillcentres were dominated by skills training in the engineering and construction industries. Both of these sectors continued to suffer severe problems during the 1980s. Within London, the rate of decline for the engineering sector as a whole was greater than in almost any other industry, and more than double the national rate of employment decline in the engineering industry (Kowarzik *et al.* 1989). Changes in the construction industry in London were equally dramatic. Between 1978 and 1985 a 7% increase in employment in this sector in Britain, contrasted with an official government figure of a 17% decline in the sector in Greater London. By 1982, 45,000 workers, more than 20% of London's construction workforce were registered as unemployed (Greater London Council 1985). Between 1981 and 1991, the construction sector in London continued to decline by around 40% (Pratt 1994b). The number of construction industry craft apprenticeships in London fell by over 35%, from 3,089 to 2,003, between 1980 and 1984.

During the 1980s, gradual policy change saw a limited increase in Skillcentre provision in London to serve the skills training needs of the inner city population. Between 1980 and 1982, the centres in north, west and south London were still training through eight existing Skillcentres and associated annexes. In east and inner south London, however, the London network had been substantially increased. Training continued at Poplar, and new centres were opened in Charlton and Deptford. Training was also re-established at Barking and Kidbrooke (later Charlton Annexe). The years 1980 to 1982 saw a very temporary peak in the provision of training through Skillcentres in London, but a peak that coincided with the MSC's announcement of a Skillcentre rationalisation programme in January 1980, almost immediately after the change of government in May 1979. Just as the inner-city issue was beginning to be addressed in terms of Skillcentre locations, the new Conservative government began a process, which would eventually move the emphasis back to the skills needs of employers and away from the social problems of the inner-city.

The change in emphasis towards inner London at the start of the 1980s, however, reflected policy decisions made under the Labour administration of 1974 to 1979, as an eventual response to a long period of industrial decline, particularly in London's Docklands. This shift in the location of provision was reinforced by the opening of a 'flagship' Skillcentre at Deptford. The national economic objectives of the MSC's earlier strategy were, by the early 1980s, largely redundant in London. Here the skills training objectives were increasingly social, in terms of relieving unemployment and 'warehousing' segments of the labour force who could not find employment with their existing skills in the midst of economic recession.

By 1984, the MSC's second Skillcentre rationalisation plan was to have significant implications for the provision of skills training through Skillcentres in London. The new Skills Training Agency had, since April 1983, been operating at arm's-length from the MSC and was required, through its business plan, to achieve full cost recovery from 1986-1987. The second rationalisation plan envisaged the closure of Skillcentre sites in

London, at Twickenham, Waddon and Waddon annexe at Sydenham, the intention being to retain training at Skillcentres in London at Barking, Deptford, Enfield, Perivale, and on a very limited basis at a new Skillcentre in Lambeth. On this occasion, Twickenham Skillcentre was reprieved but Waddon and Waddon Annexe were confirmed for closure (Greater London Training Board 1985a).

By 1985, training continued only at Enfield, Perivale, Twickenham, Deptford and Barking. In addition, the 'limited collection of courses ... known as Lambeth Skillcentre ... [and the] ... reopened (by the GLC) Charlton Training Centre ... [offered] ... no substitute for the infrastructural loss these closures represented for south London' (GLTB 1984, p. 5). Although Deptford in inner south-east London was always intended to replace some of the older existing training provision in the area, the London Skillcentre forward programme envisaged new centres in Barking and Deptford, which were opened, Camden, which was deleted from the programme, and Vauxhall, which was also removed from the programme, but which re-emerged in a limited form in Lambeth in the mid-1980s. If all these plans had been developed, it is apparent that the balance of Skillcentre sites in London would have shifted significantly towards inner London.

Within these developments, the institutional conflict which had developed over the closure of Charlton Skillcentre and the transfer of its courses to Deptford represented an important example of the failure of these agencies to develop a sense of inclusiveness around a widely held common project, to promote local business and enterprise. In April 1982, the MSC issued a consultation document proposing the merger of training provision between Charlton and Deptford Skillcentres, through the transfer of all courses from Charlton and the closure of the Charlton site. Deptford was no more than three miles from the Charlton centre, however the proposed closure generated a substantial response from within the locality (GLTB 1983).

At the instigation of the Greenwich Employment Resource Unit a series of meetings were called to discuss alternative options. These meetings were attended by representatives of seventeen interest groups; they included the Greater London Council, Manpower Services Commission, Greater London Training Board, the Inner London Education Authority, London Boroughs of Greenwich and Lewisham, Docklands Forum, Trades Councils, Adult Education Institutes and community groups from across inner southeast London (GLC 1986). In addition to these groups, a steering group was formed from the new Charlton Training Consortium (CTC) that included representatives from a number of the groups and extended membership to Woolwich College, Vietnam Trust, Lewisham Unwaged Action Group, The Simba Project, Greenwich Afro-Caribbean Association, and Greenwich Action Group on Unemployment (GLC 1986). By January 1986, the membership of the CTC consisted of representatives from over 35 local organisations. The proposed closure and transfer had provoked a response which created, for a limited period, a temporary coalition of interest groups to produce a very distinctive local training infrastructure in inner southeast London.

In this way, two very local and distinct training centres were created in London in the mid-1980s by separate institutions of labour regulation, working within the same geographical space, but with essentially conflicting interpretations of local skills training needs (GLTB 1983, 1985b). At one level and extreme, Deptford Skillcentre was the product of the MSC's *national* manpower planning strategy, and had a distinctive catchment of trainees selected to best meet those needs and objectives. At another level and extreme, and in response to the apparent inadequacy of that provision at the local level, Charlton Training Centre (at the initial instigation of the Greater London Training Board, as part of the Greater London Council) sought to respond to other *local* skills training needs. The two 'skillcentres' co-existed spatially but were directed by different agencies of labour market regulation and governance, reacting in different ways to different economic, social and political agendas.

Overall, however, labour market regulation in London in the mid-1980s had changed significantly from that of the late 1970s. The earlier commitment to skills training in inner London, a predominantly social role, was overtaken by economic objectives where Skillcentre 'cost recovery' required each individual Skillcentre to demonstrate its value to the training needs of local employers. The political shift from a 'dependency culture' to an 'enterprise culture' represented a move away from the needs of the individual towards more explicitly economic and market-led training. The new policy directives shifted attention away from the broader area-based concern of the inner-city towards a reinforced emphasis on the particular local labour market and industrial context within which each of the Skillcentres were operating. Changes in these skills training programmes in London in the mid-1980s, reflected major shifts in the national government's policy formulations regarding the means by which the state was to support industry through the provision of skilled labour. Most importantly, skills training was to be market-led and this meant the closure of 'unprofitable' Skillcentres, but an increased capacity for a flexible response to expressed needs (GLTB 1985b). These 'needs' were increasingly the needs of employers and, consequently, the training of the employed rather than the training of the unemployed. As part of this strategy, the Skillcentre review was undertaken on a centre-by-centre basis, rather than in terms of a national network. The future of each of the remaining Skillcentres was to be determined in terms of full cost recovery and the ability of each Skillcentre to sell its training services to local employers at a market price.

Change in the Skillcentre training provision in London in the 1980s reflected political change at the national level from the late 1970s and into the 1980s. The institutional support for industry envisaged under the MSC's national manpower planning strategy was played out on a national stage and was increasingly out of step with expressed local needs. Those local needs in London were increasingly the social needs of the inner-city population. Explicit social objectives, intended to reduce social unrest, were then almost immediately replaced by economic goals geared explicitly towards the local labour market, responding to employers' expressed and immediate skills training needs. In the face of increasing unemployment, economic recession and massive industrial restructuring in Britain, the London Skillcentres demonstrated institutional inconsistency of purpose, institutional inflexibility (as their training 'offer' was still dominated by declining trades), and institutional conflict (between competing agencies of labour market governance). In these circumstances, the marketisation of training services appeared to offer the potential to identify value within any particular local labour market setting.

Deregulation, Marketisation and Privatisation

In the 1980s, London experienced a fundamental shift in the structure of employment. The new model of employment structure, albeit built upon long-standing and internationally important areas of work within the London economy, rested upon the changing balance between the manufacturing and commercial and service sectors. By 1990, the service sectors accounted for 84% of all employment in Greater London. Over thirty years, the London city-region was transformed from a materials-based to a mainly information and financial service-based transactional economy (Hamilton 1991). At the beginning of the 1990s, with the effective end of the government-funded Skillcentre training provision, manufacturing and construction employment in London continued to fall by 4.2% in the year to March 1990. Although higher-level skill shortages continued to be reported, the privatisation of the Skillcentre network at the start of the 1990s, particularly within this context of persistent industrial decline in London, offered little prospect for success. This was particularly the case within a manufacturing and construction industry recession where private sector employers, who were historically reluctant to meet the costs of skills training, were experiencing high labour turnover and increasing pressures that necessitated greater labour productivity.

The privatisation of the Skillcentre network in 1990 led to the continuation of skills training at former Skillcentres at just four London sites. Buyers were not forthcoming for Twickenham and Perivale Skillcentres and they were subsequently closed. Three of the former London Skillcentre sites were sold to Astra Training Services – namely Barking, Deptford and Enfield. These three centres were subsequently closed when Astra went into receivership in 1993. The other remaining centre was in Lambeth, which was sold to Training Business Ltd.

Under this arrangement, each of the Skillcentres operated effectively as a separate cost centre, and had implicitly been prepared for privatisation since 1984 under the cost-recovery business plan of the Skills Training Agency (STA). With little scope for cross-subsidy between Skillcentre sites, and without the possibility of state funding to support social need in the absence of economic opportunity, the privatised Skillcentres were always vulnerable to employers' longstanding reticence to meet the cost of skills training. Arguably, the Skillcentres were privatised, via a market-led strategy, where no real market for skills training existed (Leonard 2001).

Following privatisation, nearly all of the former STA Skillcentres closed within a relatively short period of time. This demonstrated the limitations of the policy of transferring responsibility for industrial training to the private sector and, following the imposition of a market logic, the vulnerability of that agency/company to market failure. In December 1988, when the privatisation of the STA was first proposed by government, the national level of unemployment had been falling for two years, dropping below 2 million. The recession began in the middle of 1990 and by early 1993 unemployment was again over 3 million. The unanticipated onset of recession led directly to declining government budgets, cutbacks in private training investment, and rising local unemployment.

From this broader perspective, the privatisation of the national Skillcentre network during the 1990s was destined to fail. The private company running Skillcentres in Greater London had to operate within an unexpected national economic downturn. It sought to develop training provision within a regional economy that was subject to massive sector-by-sector industrial downsizing and restructuring in order to restore profitability. It had to sell training programmes to companies who were withdrawing from training workers in the face of economic recession. All this was occurring within the context of an increasingly neo-liberal national government policy framework that sought deregulation at the local level but a dominant controlling role at the national level (Leonard 2001).

As Peck (1992) noted, regulatory environments with their 'market-led' policy initiatives have proved vulnerable to market failure and 'a training system driven by the short-term needs of the market is self-evidently likely to produce under-investment in skill-formation' (Peck 1992, p. 343). The privatised Skillcentres had been established within a framework that demanded from them an immediate response to expressed needs. In the context of recession, those expressed needs were for employers to cut costs and labour, and not to buy in external training services. Within this context of national economic recession and cuts in public expenditure, both elements of the privatised company's income base – direct private sector contracts, and indirect public-sector funding – reduced to a point where their business was not viable. The employer-led market forces approach had to accept that in a recession, the vulnerability of such a company to market failure was an acceptable consequence of the enterprise culture and free market competition. The scope for counter-cyclical skills training and the promotion of local economic growth, particularly during periods of economic recession, was not an option for the privatised Skillcentres in London and throughout Britain.

The Skillcentre privatisation was the final change in the labour market goals, responsibilities and roles of these state-funded training centres, which had survived in London for over 60 years. This institutional persistence, coupled with inconsistency of purpose reflects the fact that these training programmes represented policy experimentation under rapidly changing economic, social and political circumstances at local and national levels. The privatised training centres were not only vulnerable to market failure, they were also constrained by the continuing post-privatisation role of central government as their principal client. The potential for institutional learning and change in order to deliver a flexible response to the expressed local needs of business was negligible given central government's concern to be seen to be responding to increasingly high levels of unemployment. Astra, even though a private company, was forced to supply traditional manual skills training to the unemployed (fulfilling social objectives) rather than new technology-based skills for the employed (responding to contemporary labour market dynamics). Ultimately, the privatised company was in competition with other training providers and firms within each local labour market. In this setting, the scope for developing institutional inclusiveness around the commonly held project of local economic development was unrealistic.

Conclusion

In the context of skills training provision within Greater London, this chapter has shown that the relationship between the state and the market is fundamental and determining. Over six decades, this relationship has taken many forms, reflecting changing regulatory needs, creating a distinctive local training infrastructure and skills training environment. The reasons for state intervention have been various, diverse and complex. Broadly, they have been *political* (to reduce social conflict and legitimate mainstream policy through both regional and inner-city policy), *social* (to counter high levels of unemployment and as part of an extended welfare and benefits programme), and *economic* (to promote and support local and national economic growth through skills training provision). These reasons are not mutually exclusive, but they all point to government intervention to counter market failure or to counter the unwillingness of private sector employers to meet the costs (social or economic) of industrial restructuring, growth and change.

From this perspective, Amin and Thrift's (1994) key determinants of institutional thickness represent an important basis for empirical analysis. However, the analysis of this chapter suggests that the role of labour market institutions in shaping and promoting local economic growth is far more complex, subtle and diverse than their a priori reasoning would suggest. These agencies have been seen to respond to a changing set of economic, social and political agendas, which have in turn operated at a variety of spatial scales, across the spectrum from the 'local' to the 'supra-national'. Institutional persistence has been easy to demonstrate. The agencies of labour market regulation in London have been reproduced in a variety of forms for over 60 years. This persistence, however, should not be confused with consistency of purpose. The example of GTCs and Skillcentres in Greater London has shown that they have performed a variety of roles, serving national, regional and local purposes, in order to promote social, economic and political goals. The same institutions and agencies of labour market regulation have served distinctly different regulatory purposes, often from the same geographical locations, but frequently in ways that have run counter to the immediate and expressed needs of local business and enterprise.

In many instances, it has been the very persistence of these labour market institutions in London that has contributed to their inflexibility and their inability to meet expressed needs and to promote local economic growth. The quasi-autonomous nature of these institutions and agencies, coupled with their changing role as demanded by the central state, has led to them frequently being out of step with local business needs. The slowness of institutional change, directed by processes of policy review by central government, has meant that inertia, lags and resistances have been common features of the operational environment of London's skills training agencies. In particular, policy initiatives have frequently been implemented from premises that had themselves been the product of earlier policies and programmes intended to serve different regulatory purposes. In these circumstances, the ability of these institutions to learn and change within the context of any particular local labour market, to support local economic growth, has been heavily circumscribed by central government and policy agendas operating at other spatial scales.

These changing agendas, often in response to labour market regulatory needs at the national level, have also run counter to the development of a sense of inclusiveness around a commonly held project, namely the promotion of local economic growth. As detailed in this chapter, institutionalised skills training, while responding to national manpower planning policy, has often failed to recognise and respond to expressed local needs. This has led to 'local' institutional conflict rather than the development of a common and widely held purpose. Even when these training agencies have been explicitly redirected to respond to the expressed and immediate needs of local business, central government has limited their activities and directed funding to achieving social rather than economic goals within a national context of economic recession and increasing unemployment.

This chapter has explored the complexity and diversity of the state-market relationship in the context of local labour market regulation in Greater London between the 1930s and the 1990s. In the changing economic, social and political circumstances of this period, particular institutions of labour market regulation and governance had been created. However, over time, their response to changing local and national, social, economic and political agendas was never consistent. As institutions, they were persistent, but not consistent in their actions towards local enterprise. These same agencies were also inflexible, demonstrating institutional inertia, lags and resistances that represent a repeated failure and inability to respond to expressed market needs in a manner that would support local economic growth. Equally, they were unable to develop a sense of inclusiveness, by developing and mobilising other agencies and organisations to support local economic growth.

Viewed from a cross-sectional, local and essentially a-temporal perspective it is quite possible that any of the training initiatives and agencies discussed in this chapter might have been interpreted as an institutional framework that sought to consolidate the local embeddedness of industry in London. The longitudinal and historical perspective presented here would suggest that, at least in the context of skills training initiatives in the Greater London labour market, the role of 'institutional thickness' and the 'processes of institutionalisation' in promoting local economic development need to be reinterpreted and reassessed.

Note

1   In broad terms, interwar regional policy had three main parts. The Industrial Transference Scheme, introduced in 1927, was designed to provide assistance for those workers in the depressed areas who sought work in the relatively more prosperous parts of the country. The Special Areas Act 1934, which was an attempt to facilitate the economic development and social improvement of those areas of concentrated, often long-term unemployment (Booth 1978). Finally, the policy of attracting work to the areas of high unemployment by creating trading estates, through legislation contained within the 1936 Special Areas (Reconstruction) Act (McCrone 1969, Booth 1982, Pratt 1994a, Mohan 1997). Interwar regional policy, under the administration of the Ministry of Labour was, however, largely about taking workers to the jobs and industrial transference remained 'the main weapon against regional unemployment' in both the 1920s and 1930s (Booth 1982).

References

Ainley, P. and Corney, M. (1990*), Training for the future:The rise and fall of the Manpower Services Commission*, Cassell, London.

Amin, A. and Thrift, N. (1994), 'Living in the global', in A. Amin and N. Thrift (eds), *Globalization, Institutions, and Regional Development in Europe*, Oxford University Press, Oxford, pp. 1-22.

Amin, A. and Thrift, N. (1995), 'Globalisation, institutional "thickness" and the local economy', in P. Healey, S. Cameron, S. Davoudi, S. Graham and A. Madini-Pour (eds), *Managing Cities: the New Urban Context*, Wiley, Chichester, pp. 91-108.

Block, F. (1994), 'The roles of the state in the economy', in N. Smelser and R. Swedberg (eds), *The Handbook of Economic Sociology*, Princeton University Press, Princeton NJ, pp. 691-710.

Booth, A. (1978), 'An administrative experiment in unemployment policy in the thirties', *Public Administration*, vol. 56, pp. 139-157.

Booth, A. (1982), 'The Second World War and the origins of modern regional policy', *Economy and Society*, vol. 11, no. 1, pp. 1-21.

Brusco, S. (1996), 'Trust, social capital and local development: some lessons from the experience of the Italian districts', in OECD (ed), *Networks of Enterprises and Local Development: Competing and Co-operating in Local Productive Systems*, LEED for OECD, Paris, pp. 115-119.

Cooke, P. and Morgan, K. (1994), 'Growth regions under duress: Renewal strategies in Baden-Württemberg and Emilia Romagna', in A. Amin and N. Thrift (eds), *Globalisation, Institutions and Regional Development in Europe*, Oxford University Press, Oxford, pp. 91-117.

Evans, B. (1992), *The Politics of the Training Market: From Manpower Services Commission to Training and Enterprise Councils*, Routledge, London.

Greater London Council (1985), *The London Industrial Strategy*, GLC, London.

Greater London Council (1986), *The London Training Legacy: The work of the GLC's Greater London Training Board*, GLC, London.

Greater London Training Board (1983), *Charlton Training Consortium – Progress Report*, TB 100, GLC, London.

Greater London Training Board (1984), *MSC Proposals to Close Skillcentres*, TB 317, GLC, London.

Greater London Training Board (1985a), *Skillcentre Closures – Update Report*, TB 418, GLC, London.

Greater London Training Board (1985b), *The Skillcentre Training Agency and Adult Training*, TB 332, GLC, London.

Hamilton, F. (1991), 'A new geography of London's manufacturing', in K. Hoggart and D. Green (eds), *London: A New Metropolitan Geography*, Arnold, London, pp. 51-78.

Kowarzik, U., Williamson, E. and Leonard, S. (1989), *London Labour Market Review 89*, Employment and Training Group, London Research Centre, London.

Leonard, S. (1999), *Geographies of Labour Market Regulation: Industrial Training in Government Training Centres and Skillcentres in Britain and London 1917-93*, unpublished PhD thesis, Department of Geography and Environment, London School of Economics and Political Science, University of London.

Leonard, S. (2001), 'Regulating labour, transforming human capital and promoting local economic growth: The case of the UK Skillcentres initiative', in D. Felsenstein and M. Taylor (eds), *Promoting Local Growth: Process, Practice and Policy*, Ashgate, Aldershot, pp. 347-364.

Leonard, S., Maginn, A. and Kowarzik, U. (1991a), *Where Have All the Jobs Gone? Skills Implications of Employment Change in London 1980-1990*, Employment and Training Group, London Research Centre/Association of London Authorities, London.

Leonard, S., Maginn, A. and Williamson, E. (1991b), *London Labour Market Review 90*, Employment and Training Group, London Research Centre, London.

MacLeod, G. (1997), '"Institutional thickness" and industrial governance in Lowland Scotland', *Area*, vol. 29, no. 4, pp. 299-311.

Martin, J. (1966), *Greater London: An Industrial Geography*, Bell, London.

Ministry of Labour (1931), *Report of the Ministry of Labour for the Year 1930*, Cmnd. 3859, HMSO, London.

Ministry of Labour (1947), 'Vocational training scheme: recent developments', *Ministry of Labour Gazette*, p. 46.

McCrone, G. (1969), *Regional Policy in Britain*, Allen and Unwin, London.

Mohan, J. (1997), 'Neglected roots of regionalism? The Commissioners for the Special Areas and Grants to Hospital Services in the 1930s', *Social History of Medicine*, vol. 10, pp. 243-262.

Peck, J. (1992), 'TECs and the local politics of training', *Political Geography*, vol. 11, no. 4, pp. 335-354.

Pratt, A. (1994a), *Uneven Re-production: Industry, Space and Society*, Pergamon Press, Oxford.

Pratt, A. (1994b), 'Industry and employment in London', in J. Simmie (ed), *Planning London*, UCL Press, London, pp. 19-40.

Sheldrake, J. and Vickerstaff, S. (1987), *The History of Industrial Training in Britain*, Avebury, Aldershot.

Chapter 12

Diasporic Embeddedness and Asian Women Entrepreneurs in the UK

Irene Hardill, Parvati Raghuram and Adam Strange

Introduction

The literature on entrepreneurial embeddedness has highlighted the ways in which social relations underlie and reinforce production relations within geographically bounded spaces (Granovetter 1985). It is argued that the strength of such relations facilitates entrepreneurial success in regions, most notably with the development of the idea of industrial districts[1], as exemplified in Emilia-Romagna in northeast Italy (Harrison 1992, Piore and Sabel 1984). However, much of this literature has focused on the ways in which these businesses have utilised previously existing social relations which are constituted and reinforced by spatial proximity largely in 'case study' industrial districts in advanced capitalist economies (Putnam 1993, Salais and Storper, 1993). In essence, the socio-economic milieu[2] provides the 'organising context' for entrepreneurial activities (Leipitz 1993). These studies have focused on groups of people with a deep attachment to their places of production, and this attachment is reflected in their rootedness to place. Thus, this work has focused on relatively 'unshifting populations'. Lack of movement is deeply implied in embeddedness theories (Hardill *et al.* 1995).

In this chapter we extend this discussion by examining the ways in which embeddedness operates among minority ethnic firms in the English Midlands region of the UK[3]. These firms have been established and are being run by women from families with complex migration trajectories. The families have migrated often more than once, and hence have few historical linkages with the places where their businesses have been

207

established or indeed trade. Such businesses are marked by the (social and spatial) mobility of their owners, yet the ethos and the material conditions of production in these businesses are underlain by strong ties to their 'place of origin', their diasporic home. Furthermore, the success of these businesses is dependent on social relations established with producers and consumers within the diaspora. Linkages with place are central to these businesses. Hence these businesses are embedded not at the local scale, but at the global scale. These links become realised through a number of practices: common consumption practices because of these links; production chains which link raw material, know-how, production processes and the market. Common to these links is not the long established spatially bound social relations they involve, but their ability to form new social relations on the common knowledge of being part of one diasporic formation. Spatial relations are not locally conjunctural as in the classical embeddedness literature, but globally positioned through imagined relations with a locale, a place, the diasporic 'home'.

After this introduction, the chapter is divided into six sections. The next section, Section 2, examines the notion of embeddedness within existing literature on minority ethnic entrepreneurship. Section 3 outlines the context of the case study. Section 4 outlines some of the specificities of the market niche that we focus on: Asian[4] women's fashion clothing, and the following two sections look at social and spatial relations utilised by our case study firms at two different spatial scales: the global and the local.

Embeddedness and Minority Ethnic Enterprise

Within the growing body of academic and policy literature on minority ethnic enterprise, two aspects of 'embeddedness' are of particular significance. The first of these is the importance of social relations in business formation and operation. Thus Phizacklea and Ram (1996, p. 320) note that, 'most of the minority ethnic literature acknowledges the extent of social embeddedness in the raising of start-up capital and the provision of labour and custom for the minority ethnic firm through kinship and co-ethnic networks'.

However, it is argued that the ability to form and/or utilise such networks for entrepreneurial benefit is not consistent across all ethnic groups. A key theme of the minority ethnic business literature has been the explanation of differential levels of business formation and 'success' exhibited by different minority ethnic communities. In the UK, this has translated into frequent comparisons of the very different experiences of more active (and more highly visible) South Asian[5] groups, with self-employment rates well above the national average, and African-Caribbeans, who are much less active. Central to explanations of this 'Afro-Asian gap' (Barrett *et al.* 1996, p. 788) are the higher levels of social embeddedness believed to be enjoyed by the UK's South Asian communities. These provide a '"cushion" of co-ethnic support' (Ram *et al.* 1997, p. 90) throughout the business, but perhaps most significantly in the core areas of raising capital (Jones *et al.* 1994) and recruiting trusted and affordable workers (Phizacklea and Ram 1996), often including unpaid family labour (Storey 1994). Conservative accounts hold up the 'strong social ties', which carry within them a collapse of other conservative mantras – 'family values' and 'entrepreneurial values' (Kofman *et al.* 2000). Thus, the Head of Small Business Services at a major UK clearing bank remarked that, '[in] general these [Asians] have proved to be low risk communities as far as the bank is concerned. This is primarily due to the very strong family ethos' (Global Consulting 1996, section 3, p. 18). Yet this dependence on social networks is often a result of the exclusionary processes of mainstream business support agencies, and even more importantly the funding agencies, particularly the banks (Jones *et al.* 1994).

In addition, the literature talks of the importance of 'minority ethnic niches' and 'minority ethnic enclaves'. Minority ethnic niche markets are:

> ... specialised fields of demand ideally adapted to the cultural and business practices of ethnic firms ... in which they enjoy a competitive edge, even a monopoly advantage over non-minority firms ... perhaps the most obvious of such niches is that provided by the customer potential of the ethnic minority population itself. (Ram and Jones 1998, p. 35)

Examples, therefore, might include the production and/or distribution of specialist foods or fashions. Minority ethnic entrepreneurial niches[6] are characterised by their concentration in low-income, high labour intensity industries, such as hairdressing among African-Caribbeans and grocery stores amongst Asians. Such niche markets are characterised through the formation of enclaves, where distribution outlets, such as specialist retail shops, are clustered. 'Ethnic entrepreneurial niches are industrial clusters of co-ethnic entrepreneurs at levels far exceeding chance' (Razin and Light 1998, p. 557). The development of a number of sizeable Asian business enclaves within the UK is facilitated by the spatial concentration of the population in a limited number of cities across the country (including London, Manchester, Bradford and Leicester), and in a limited number of residential neighbourhoods within those cities (Ram and Jones 1998). Hardill (1998) has highlighted the relationship between size of ethnic group in an area and the size of a retailing cluster. Thus, the presence of significant numbers of co-ethnic customers supports the growth of niche markets which, in turn, take on a physical form in the development of local enclaves.

Implicit in much of the discussion of minority ethnic business and embeddedness, therefore, is the idea of social and economic relations conducted in spatial proximity – of networks within a discrete geographical space. This reflects the literature on the embeddedness of smaller firms more generally, as exemplified by examinations of the industrial district thesis in northeast Italy and elsewhere (see Curran and Blackburn 1994). In the literature relating to UK minority ethnic businesses, however, these spatial linkages are couched in terms of the spatial concentration of retailing outlets, and there is little or no discussion of the ways in which social relations are used in production.

The significance of such linkages is, however, increasingly being realised. A number of speakers at the 1996 Ethnic Minority Businesses National Conference in Leicester[7] commented on the need to:

> ... understand and appreciate the opportunities that the local ethnic minority businesses have in making trading links with the countries of their origins or their parents' origin. (Karl Oxford, African-Caribbean Economic Establishment, Bradford; Global Consulting 1996, section 3, p. 34)

They recognised that they had, 'an understanding of the areas and regions that we come from and we understand the kind of attitude and behavioural patterns' (Bob Clarke, Development Manager, The Global Trade Centre, North London; Global Consulting 1996, section 4, p. 41). They also highlighted the need to enter international trade formally in order to gain recognition (Uday Dholakia, Senior Partner, Global Consulting; Global Consulting 1996, section 4, p. 41) as minority ethnic firms in a competitive environment. These ideas were perhaps best summarised by Pen Kent, Executive Director, Finance and Industry, Bank of England:

> Now what are the strengths? Have you got some strengths as well as a few handicaps? You know you have and one of them is 'Global Links' which was mentioned yesterday. Has the African-Caribbean Community got some links

which gives you a start point, an entry point to import, to export, to build, to create, to do things etc.? Has the Asian community got those? I believe and know the answer is yes because I've met some of the people who do actually use their Global Links. They are often like a small inheritance of your own family histories and those can be built upon. (Global Consulting, 1996, section 8, p. 9)

In the rest of this chapter we examine the ways in which four Asian businesswomen utilised the embedded social relations inherited through their diasporic histories to establish and run successful businesses.

The Study

This chapter draws upon findings from a study of minority ethnic entrepreneurs in the East Midlands region of the UK[8]. Early in the study, we identified the city of Leicester as the key site for our research within the region. As a location for a study of minority ethnic business the choice of Leicester is a significant one, most notably with regard to the city's relatively large Asian community. Leicester has an ethnically-diverse population, with 71.5% identifying themselves as White, 22.3% Indian, 1.0% Pakistani, 1.5% Black Caribbean and 0.3% Black African in the 1991 Census of Population (OPCS 1992, Table 6).

In view of the dominance of Asian entrepreneurs and 'entrepreneurial success stories' in Leicester and their high visibility, our study concentrated on two groups whom we identified as minorities within the minority: African-Caribbean entrepreneurs and Asian women entrepreneurs. In this chapter we draw upon interviews conducted in the second part of the study[9].

The word 'Asian' was used as a collective term, to differentiate people who originated from the Indian subcontinent from the African and European populations in East Africa (Burja 1992). Subsequent to the migration of large numbers of these 'Asians' to the UK, following the independence of many African states in the 1960s and the policies of Africanisation that followed, the term has been transposed to the UK context. It has come to include all populations originating from the Asian subcontinent and now resident in the UK, whether they migrated here from East Africa or directly from the subcontinent. The Asian population – whether the result of direct or indirect migration – is very diverse, being marked by linguistic, religious, regional and sectarian differences. Furthermore, the 1980s and 1990s have also witnessed a growing social polarisation between and within Asian communities, with those who were indirect migrants emerging as real 'winners'. New divisions are emerging therefore within Asian communities, in terms, for example, of educational attainment, professionalisation and entrepreneurial activity (Metcalf *et al.* 1996). This polarisation is further illustrated by the fact that the occupations with the highest proportion of minority ethnic workers are plant and machine operators and the professional occupations (Phillips and Sarre 1995).

Many of these distinctions are reflected in Leicester. Here the Asian population is concentrated in the inner city, particularly in the Highfields area immediately to the east of the city centre, which over the last century became the first home of successive waves of migrants (Nash and Reeder 1993). With the arrival of labour migrants from India and Pakistan to meet the post-war labour shortages in the 1950s and early 1960s, the family reunion migrations that followed, and the more substantial migrations from East Africa in the late 1960s and 1970s, Highfields has become the home for a large number of Asians. East African Asians eventually settled in the Belgrave area because of a shortage of accommodation in Highfields. They found in Belgrave an inner city area that was depopulating, with vacant residential and business premises. Thus, with the coming of the East African Asians, 'Belgrave [to the north of the city centre] was transformed into a thriving residential and shopping area, with a lively business community' (Nash and Reeder 1993, p. 27).

Since that time the Asian communities of Leicester have continued to grow in number. The 1991 Census of Population highlighted the breakdown of concentration of the Belgrave ward as a place of East African Asian residence (but not of enterprise) (O'Connor 1995), as a result of suburbanisation (Byrne 1998). Retailing is concentrated in this ward, particularly along the Belgrave and Melton Roads, which runs out of the city to the north-east, and which has also become the centre for annual Diwali[10] celebrations in the city and the region. Significant Asian retail centres also exist on other major routes into the city, notably on Narborough Road, which approaches the city from the southwest, where a number of wholesale outlets are interspersed amongst the retail shops. The Belgrave shopping area on Belgrave and Melton Roads, however, represents the most important retail centre for Asians in the East Midlands.

In our study, we identified our first interviewee through our existing contacts, and subsequent respondents were identified by the 'snowball' method. In this paper we draw upon four of these detailed interviews with 'Rama', 'Malini', 'Indira' and 'Kala' (Table 12.1). All are East African Asian women in their thirties who come from entrepreneurial family backgrounds, but have subsequently identified their own business opportunities in the fashion sector. Interviews were accompanied by self-completion questionnaires and lasted between one and three hours. The interviews were taped and transcribed and the interviewees were sent copies of the transcripts to comment upon.

Table 12.1   Case study Asian women entrepreneurs

Malini (Company 1) is in her early 30s. Her family migrated from Gujarat, India to Zambia in 1948, where they retailed children's clothing. Her father migrated to the English Midlands in 1954, being joined by his wife and two children in 1958. Malini was born in the Midlands where her parents owned a grocery shop. She wanted to go into business, but no family money was forthcoming, 'not a penny, my mother wouldn't give it to me, not a single penny has come from my family. She didn't approve of me going into business initially because it's the usual scenario you know, you have graduated and you should get married. Parents always want to give everything to their boys, the girls have to struggle'. But eight years ago she joined the wife and son of the founder of Company 1. All three are East African Asians. The business, 'went limited', and thanks to Malini the clothing retail business was transformed. It now, 'sells Indian clothes, but they are modern stuff, as well as traditional'. Her male business partner Rajiv, 'always feels that I am on a female mission and he is very supporting'. They own two other retail outlets in dynamic South Asian retail centres in London. Their garments are manufactured in India, 'about eighty per cent [is for the "British market"]...besides our three retail outlets we sell to other people...and export to Mauritius, South Africa and America'. They employ 41 in the UK and seven in an office in India.

Kala (Company 2) is in her late 30s. Her family (parents, elder brother and she) left Kenya in 1973 and settled in London. Her father first worked for an insurance company and her mother in a factory, but then they acquired a newsagent's shop. Her in-laws (from Zambia) began trading in 1983, but, 'once I joined in 1984, bringing fresh ideas and new things to do ... silk shoes, salwaar kameez ... things increased ... now we have bigger premises'. The readymade garments sold in their Leicester shop are made in Bombay, India and the owners make regular trips to India as they are involved in the, 'designs and choice of material'. They employ four people in their Leicester shop plus about six people elsewhere. Ninety per cent of their customers are Asian, especially 'Zambian Asians', many of whom they know personally.

Indira (Company 3) is 40 years old. Her family (parents and elder brother) left Kenya in 1975, where they had a retail outlet. Her family by marriage established the shop in 1991 in Leicester and retails, 'clothing – exclusive wear. There was a market for Asian designer wear ... we established the shop six years ago because nobody was doing that [designer wear] for Asian ladies'. The shop, 'it's a boutique ... ladies, gents, childrens, exclusive'. Her father-in-law, husband and herself own the business. Indira had the idea for the shop, 'my husband supported me all the way'. The Leicester shop employs four women and they also have a second fashion retail shop in Birmingham. About sixty per cent of the stock for their shops comes from India and Pakistan, and the majority of their customers are Asian. When asked about plans for the future of the business, it was, 'expansion and looking forward [to opening] more outlets'.

Rama (Company 4) is in her 30s, and her family migrated from Kenya, where they had lived since the 1920s. Her grandfather had established a business in Kenya in the 1920s and her father and his two brothers ran the business, which comprised a grocery shop, a crockery shop and a printing press. Her family moved to the UK in 1979. She is still single, 'I still live with my parents and as they are my parents I love to do things for them'. Company 4 is a sole proprietorship, which was opened by Rama. Like Malini, Rama had a business idea, 'my parents and other family members have supported me [they] all thought I was mad. I have had moral support from friends and family but to start with my own family thought "oh it's a whim she's going through", but it's been my dream and I worked damned hard to make that come true'. Her parents and other members of her family supported – morally not financially – her decision to open her designer-wear shop. But, before this she, 'used to be a hairdresser ... when Asian girls didn't do hairdressing. I worked for someone else ... then I opened my own salon. Then I started studying again ... I had always had a passion for fabrics'. She gets her fabrics from India, 'I physically go there, I buy my own fabrics, I get embroiderers ... and garments stitched to my specifications ... in Rajasthan, Gujarat, Delhi, Bombay and in the South'.

Source: Minority Ethnic Business Survey, 1996-1997.

The Fashion Industry

Gender, class and ethnicity operate to co-produce different forms of clothing consumption patterns. Although there are large variations within clothing patterns in the Indian subcontinent, there has been a tendency for certain garments, such as the sari and the salwaar kameez, to gain a hegemonic status, both in the subcontinent and in its diasporic populations (Tarlo 1996)[11]. Hence, the fashion garment industry has concentrated on these two forms of clothing, or variants of them. Asian men are, on the whole, less likely to wear 'Asian clothes' every day, as indeed is the custom and practice in the Indian subcontinent. They only wear 'Indian' clothes for such special occasions as their own wedding. Asian women, on the other hand, are much more likely to wear 'Asian clothing' every day, and it is unlikely that they would wear 'Western clothing' for *any* special occasion – for annual festivals or for social gatherings surrounding lifecourse events, such as birth, marriage or death.

These generalised patterns, of course, do vary with religious affiliation and region of origin in the Indian subcontinent, as well as area of residence in the UK. Thus, for example, men appear to be more likely to wear a salwaar kameez in areas where there is a relatively high degree of segregation, or when visiting such areas. On the whole, Muslim men are more likely to a wear salwaar kameez than Hindu men. Finally, an increasingly important form of differentiation is emerging with the growth in the number of the Asian service class of managers and professionals. Social stratification within British society is also reflected (and perhaps even magnified) within the Asian society, leading to specific forms of consumption. It is this group of 'conspicuous consumers', who cut across cultural, racial and ethnic communities, who have been targeted by the group of entrepreneurs highlighted in the empirical evidence which follows.

The four entrepreneurs we interviewed all purchased or produced their products in India for this retail market. Malini, for instance, is closely involved in the design of the garments, using designers in London and in India, and actively uses diasporic connections for organising the enterprise: for the purchase of the raw materials, for the organisation of garment production by subcontractors in India, and for marketing. Subcontractors, who manufacture her designs to her specifications, make the garments that Malini sells. This takes place in Cut-Make-and-Trim (CMT)[12] units in India, because she feels that the craft skills required for the production of the clothes are only available in India, but also because of labour costs. Malini thus uses her knowledge of the market she is exploiting through her inputs into garment design. She orders exclusive limited runs, and changes the stock every month. Similarly, Kala characterises her customer as, 'somebody that is looking for exclusive goods', and stresses the organisational and geographical flexibility that is required in sourcing appropriate products for this discerning consumer. Stock comes primarily from India (80%), but also from Japan, to designs drawn up both in India and the UK. And getting the right combination of materials and design necessitates extensive travel, visiting suppliers in India. Of her suppliers, therefore, she acknowledges the importance of, 'working in conjunction with each other, because I know my clientele would like something like this'.

Fashion retailers, unlike other clothing retailers, closely control their production chain. As sales are linked to measures of quality and to exclusivity, the entrepreneurs we interviewed were closely involved in all stages of the production process. Crewe (1996) notes, therefore, that in the

British fashion industry proximity to the consumer is now of increasing importance to the producer, but with minority ethnic businesses this is not the case. They have been able to take advantage of global communication networks as well as their diasporic connections to separate production from consumption. They can take advantage of the cheap labour and the craft skills available in India to produce the goods, but still sell at First World prices. Malini, for instance, overcomes the problem of distance by using air transport to move garments to the UK. These entrepreneurs provide us with an example of the ways in which globalisation has been actualised in everyday practices, with heterogeneity incorporated within these narratives of globalisation. This heterogenisation is embedded in relations of power and control and decision-making within the textile and clothing filière[13] between the UK (retail outlets) and India (textile manufacture and garment production) and differential wage rates, but is also embedded in ideas of a spatially located cultural capital, i.e. skills.

In addition to her retail outlets, Malini also sells to retailers in other urban centres in the UK with smaller Asian populations. The company also exports to countries with significant Asian populations, notably South Africa and the USA. This is done through a parallel export-import firm, which Malini has established. For the US market she uses an Asian American, whose parents live in the UK. The American client, during business trips to the UK that also become family trips, undertakes quality checks and ordering. Thus, she actively uses diasporic connections to produce and sell an image, but she also recreates the diasporic culture through her production and marketing strategies. In addition, she makes skilful use of technology, as she advertises on Zee TV, a television channel that is produced in Mumbai (Bombay) but is available to cable television owners – her differentiated clientele – in the diaspora. Malini thus uses global communications to sell to global markets. She also has regular features in newspapers such as the Eastern Eye and the Buzz, because, 'they do a lot of international advertising' (Malini).

The spatiality of consumption and production in the retailing of Asian clothing operates at a number of scales, and in the next two sections of the chapter we focus on two spatial scales around which embedded production-consumption relations appear to revolve[14].

Global Connections: Spatial Distance and Social Proximity

Although all the entrepreneurs we interviewed had some family background in entrepreneurship, the businesses that they currently owned were either new or markedly different from those with which they had earlier been linked. Rama's father owned businesses in Kenya – a grocery shop, a crockery shop and a printing press. Malini's father owned a grocery shop in the UK, Indira's parents had a retail outlet in Kenya, and Kala's parents were newsagents. Thus, Kala, when asked if her business had been established in Zambia and had been transported to the UK, replied:

> Not really because [it is] a completely new country here. So you really do have to start it off again. You knew the business in the sense that you were in the sari trade. Over there they have a lot of wool and knitting wear. All kinds of things. It was more orientated to textiles ... Once I joined in '84, bringing fresh ideas and new things to do. And as trends came in we either choose silk shoes, silk handbags, salwaar kameez. We had ready-made blouses for saris. Things increased. (Kala)

Hence, even where firms had been inherited or developed from pre-existing businesses, the shift to retailing fashion garments involved establishing new production linkages:

> Because these are very exclusive fashion garments, we work with a lot of different suppliers, depending on the type of item that we want. The handmade or the woven saris we actually go to the weavers itself, to the places that actually weave. So we have to travel quite extensively in India to find different things. (Kala)

However, existing relations were central to establishing these new relationships:

> Q. When you thought you wanted to expand it particularly, did you seek any help?
> A. Well, mainly the major support we had was from our suppliers in India. They have been the major source of moral support or financial support, all kinds of support really. Giving the ideas of all the new fashions that are available. Giving us a contact to other people. They have been exporting all our goods. They were exporting in Zambia as well. To the family. Then even here. They were the ones to suggest that if we were going to move, Leicester is a good place. Because they'd already been in this country, actually exporting to other people as well. So it was a

relation more of not just a business relationship, it was more a family
relation in that sense. (Kala)

These links are particularly important in the fashion industry, where
product and design quality are important criteria in identifying the products
for sale. The problem of distance between producers and retailers may be
partly overcome where relationships of trust have been established:

> When you are running a business there are many other problems, not just
> getting the goods at the right time and the right choice. Because if you think
> alike it's the only way you can rely on them to pick up the phone and they say
> 'we are sending you 25 of these'. And you know that there will be something
> you would pick yourself. So those connections really helps. (Kala)

Good social relationships with the producers can be particularly
important for women entrepreneurs:

> There has to be support from there, otherwise it cannot run. Everything
> depends on importing the goods. Again this is where being a lady comes in,
> because you cannot be just upping and going as often as you would like to. Or
> as often as the business requires to. This is where support comes in. However
> much I would like to go for a buying trip, I have to really organise the
> schedule well before I can leave. (Kala)

At the same time women may find it harder to establish such close
relations, especially when they are not mediated through long-standing
personal or family relations. Thus, when Indira was asked whether she
knew any of the people in India that she does business with personally, she
replied:

> No, none of them are my relations and I had to fight through that. India is a
> very difficult place to trade with and once they know you are a foreigner, and
> a woman at that, they are still searching for my credibility, even now, after all
> these years. (Indira)

Despite the significant clustering of Asians in Leicester, our firms had
greater production links with India than with the 'enclave' within which
they were physically located. All firms sourced 80% or more of their raw
materials from India, with some design input from Indian designers as well
as being manufactured there, albeit under strict supervision from the
entrepreneurs themselves. Thus, Rama said:

I physically go to there [to India], I buy my own fabrics, I get embroiderers ... and garments stitched to my specifications. (Rama)

And, from Kala:

Q. You don't actually import the raw materials?
A. Everything is chosen, made up and brought there.
Q. What about the actual production ... is it all brought to a factory in Bombay?
A. It depends, really. Certain things happen in Bombay, certain processes happen outside Bombay. It's basically in India. (Kala)

In contrast, there are few linkages between the firms located within Belgrave and Melton Roads. Indeed, as Indira pointed out:

... to tell you the truth, there are very very few people on this one mile that have been into my shop. (Indira)

Local Embeddedness?

The locational factors influencing minority ethnic enterprise are specific. Areas such as Belgrave and Melton Roads in Leicester have become the centres for commerce as the total market is too small for entrepreneurs to move out of such economically determined locations. As was noted earlier, second and third generation Asians (especially Indians and East African Asians) have entered the professions (Metcalf *et al.* 1996), achieving social mobility and creating a middle class. This influences the strategies of successful retailers, as such a middle class living in large urban centres with high minority ethnic populations offers the only economically viable locations for specialist retailers. For instance, Malini feels that:

... in this country there are only six major areas that you can retail Indian products effectively, and that's obviously East London, Wembley, Southall – that's the three major areas covered there – Birmingham, Leicester, Manchester. Everywhere else there are Indian areas, but they are just too small. (Malini)

She feels that everyone travels to these centres because, 'England is not very big –how long does it take you?' Viable enterprises, therefore, have particular attachments to place. At the same time, the convergence of niche shops is also based on the need to be located in a particular ethnic and class milieu. As Malini is selling a certain brand of ethnicity, she needs to locate in a vibrant 'Asian consumption site', adjacent to other successful, exclusive Asian shops, such as jewellers and good quality restaurants. The shop is located in the centre of the shopping area, 'down in the middle, [where] you can't miss it.' It benefits not only from its location in the street, but also from its relative location in that part of the street where the more exclusive shops with higher turnovers are located. Malini's shop thus has a firm attachment to place.

The locational strategies of the firms we studied also suggest that such shops locate in centres that have become identified with particular forms of consumption. These centres attract those who visit such consumption sites not merely to purchase 'goods', but also to consume a particular social environment. Practices such as dressing in Indian clothes to 'go shopping' on these streets indicate that people 'go shopping' not merely to buy, or even to see, but also to be seen. Thus, they not only consume the milieu, they also help to create it by articulating their version of a group identity through their clothing. Retail centres are thus spaces of consumption as well as performative spaces which not only reproduce but also create some sense of shared identity. A visit to Belgrave and Melton Roads not only involves shopping for Asian fashion clothing, but also provides a locale in which these clothes may be worn. Hence, the locational strategies of firms are reinforced as they become embedded in particular places.

At the same time, the disparities within these streets, both in the nature of the shops and the nature of the clientele, question the cohesiveness of the 'social milieu' being created and consumed, and of the identity being expressed. The diasporic sphere of retailing assumes an homogeneity of experiences based on place of origin as well as migratory experiences, but as this study shows, both are fractured along lines of gender and class, among a host of other social variables.

Conclusion

In this chapter we have sought to contribute to the literature on embeddedness by extending the spatial scales within which embeddedness operates. The businesses that we studied employ production relations, which are stretched across space. Where these relations are also social relations, entrepreneurs find this a distinct advantage, particularly if they are women. Yet not all such relations are social, and the entrepreneurs have forged new production and consumption links to establish successful businesses. Hence the embeddedness of social relations, while a distinct advantage, is not always of spatially proximate people. Embeddedness then lies in imagined relations with home, and the realisation of these in consumption practices. Moreover, the embeddedness involves not only production practices, but also consumption practices where links between retailers and consumers are forged through common histories, through similar migration patterns and experiences of hybridisation. At the same time, the local scale is relevant to these businesses, as the size of the minority ethnic populations in the locality and their social and economic well being influences the viability of the enterprise.

Finally, we have contributed to the literature on minority ethnic enterprise by highlighting the positive ways in which embeddedness may be utilised by firms. Most of the existing literature on spatial fixity and sectoral specialisation focuses on the need for break-out, but as in the wider fashion industry, super-specialisation and identifying smaller but dynamic niches (with high margins) can also work as an effective entrepreneurial strategy.

Acknowledgments

The article draws on research funded by Nottingham Trent University. David Graham, Irene Hardill, Parvati Raghuram and Adam Strange undertook the research. We acknowledge the research assistance of Sameera Mian and Reena Mehta, and are indebted to the four interviewees. The views expressed here are those of the authors alone.

Notes

1   The term 'industrial district' was first used by Alfred Marshall when writing over a hundred years ago about contemporary industrial development in Sheffield and Lancashire (UK). Marshall felt an industrial district – an agglomeration of SMEs in the same area – could be a mode for organising production in certain manufacturing industries alternative to the large firm (Marshall, 1920, 1923).

2   The milieu is said to group together in a coherent whole a production system, a culture and actors (Maillat *et al.* 1993). The coherence between the various actors lies in their common approach to situations, problems and opportunities. The milieu, as a framework of analysis, defines an interior and an exterior, as the processes lead to differentiation between the milieu and other entities around it (Hardill *et al.* 1995).

3   Raghuram and Strange (2000) discussed some of the ways in which discourses construct and obstruct entrepreneurship, particularly in the context of minority ethnic groups, and in consonance with this position we have used the term 'minority ethnic' in this paper, rather than the more commonly used term 'ethnic minority' (except in quotations).

4   The term Asian was a collective term used to differentiate people who originated from the Indian subcontinent from the African and European populations in East Africa

(Burja 1992). It has now been transposed to define all people originating from the Indian subcontinent, whether migrating directly or indirectly to the UK. We use this transposition in this chapter.

5   Within the South Asian communities rates of self-employment do vary, high levels of self-employment are recorded especially by East African men and women (Owen 1995).

6   Such studies rarely discuss the ways in which stereotypical images of minority groups direct people from certain groups to certain types of activity, by allocating or withdrawing start-up funding. Rather they portray sectoral niches as products of natural proclivities or of a lack of imagination. Breakout from ethnic niches becomes the key answer to success (Global Consulting 1996).

7   The Ethnic Minority Businesses National Conference was held in Leicester from 27-28 March, 1996. We would like to acknowledge the support of Global Consulting, who provided us with the Conference Report and Best Practice Guide from which these quotes are drawn.

8   The project was funded by Nottingham Trent University's Research Enhancement Fund, and we would like to acknowledge the other member of the project team, David Graham.

9   We have focused on the specific gender dimensions of this entrepreneurship elsewhere (Raghuram and Hardill 1998, Hardill and Raghuram 1998).

10   Diwali is the Hindu Festival of Light.

11   The salwaar kameez is a dress composed of two garments: a long shirt and loose trousers. Different styles are worn by men and women. Both in India and the Indian diaspora there has been a growing tendency to replace the sari with salwaar kameez, even for formal occasions. Hence, a market for high quality salwaar kameez has been created. While this market is also growing in India, it is already saturated and this was recognised by our interviewees.

12   CMT units are subcontactors who manufacture garments to the specifications of the contractor, who provides the materials.

13   A filière is a series of economically and technically inter-related operations placed between the availability of the raw material and that of the finished product (Maillat 1982).

14   While these scales are used to develop the argument, we would like to reiterate that they are interdependent and interlinked, and not hierarchical.

References

Barrett, G.A., Jones, T.P. and McEvoy, D. (1996), 'Ethnic minority business: theoretical discourse in Britain and North America', *Urban Studies*, vol. 33, no. 4-5, pp.783-809.

Bujra, R. (1992), 'Ethnicity and class: the case of the East African 'Asians'', in T. Allen and A. Thomas (eds), *Poverty and Development in the 1990s*, Open University Press, Milton Keynes, pp. 437-461.

Byrne, D. (1998), 'Class and ethnicity in complex cities: Leicester and Bradford', *Environment and Planning A*, vol. 30, pp. 703-720.

Crewe, L. (1996), 'Material culture: embedded firms, organisational networks and the local economic development of a fashion quarter', *Regional Studies*, vol. 30, no. 3, pp. 257-272.

Curran, J. and Blackburn, R. (1994), *Small Firms And Local Economic Networks*, Paul Chapman, London.

Global Consulting (1996), *Ethnic Minority Businesses National Conference: Conference Report and Best Practice Guide*, Global Consulting, Leicester.

Granovetter, M. (1985), 'Economic action and social structure: a theory of embeddedness', *American Journal of Sociology*, vol. 91, pp. 481-510.

Hardill, I., Fletcher, D. and Montagné-Villette, S. (1995), '"Small firms", "distinctive capabilities" and the socio-economic milieu: Findings from case studies in Le Choletais (France) and the East Midlands (UK)', *Entrepreneurship and Regional Development*, vol. 7, no. 2, pp. 167-186.

Hardill, I. (1998), *Diasporic Business Connections: An Examination of an Asian Business District*, Paper presented at the Economic Geography Research Group of the RGS/IBG Conference, Manchester.

Hardill, I. and Raghuram, P. (1998), 'Diasporic connections: Case studies of Asian women in business', *Area*, vol. 30, no. 3, pp. 255-261.

Harrison, B. (1992), 'Industrial districts: old wine in new bottles', *Regional Studies*, vol. 26, pp. 469-483.

Jones, T., McEvoy, D. and Barrett, G. (1994) 'Raising capital for the ethnic minority small firm', in A. Hughes and D. Storey (eds), *Finance and the Small Firm*, Routledge, London, pp. 145-181.

Kofman, E., Phizacklea, A., Raghuram, P. and Sales, R. (2000), *Gender and International Migration in Europe: Employment, Welfare and Politics*, Routledge, London.

Leipitz, A. (1993), 'The local and the global: regional individuality or interregionalism', *Transactions of the Institute of British Geographers, New Series*, vol. 18, pp. 8-18.

Maillat, D. (1982), *Technology: Key Factor for Regional Development*, Georg, Germany.

Maillat, D., Crevoisier, O. and Lecoq, B. (1993), 'Réseau d'innovation et dynamique territoriale: le cas de l'arc jurassien', in D. Maillat, O. Crevoisier and L. Senn (eds), *Réseau d'Innovation et Milieux Innovateurs: Un Pari Pour Le Développement Régional*, EDES, Neuchâtel, pp. 17-50.

Marshall, A. (1920) (reprint 1959), *Principles of Economics*, Macmillan, London.

Marshall, A. (1923), *Industry and Trade*, Macmillan, London.

Metcalf, H., Modood, T. and Virdee, S. (1996), *Asian Self-Employment: The Interaction of Culture and Economics in England*, Policy Studies Institute, London.

Nash, D. and Reeder, D. (1993), *Leicester in the Twentieth Century*, Alan Sutton Publishing and Leicester City Council, Leicester.

O'Connor, H. (1995), *The Spatial Distribution of Ethnic Minority Communities in Leicester, 1971, 1981 and 1991: Analysis and Intrepretation*, Centre for Urban History and the Ethnicity Research Centre, University of Leicester, Leicester.

Office of Population Censuses and Surveys (OPCS) (1992), *1991 Census County Report: Leicestershire (Part 1)*, HMSO, London.

Owen, D.W. (1995), *Ethnic Minority Women and Employment*, Equal Opportunities Commission, Manchester.

Phillips, D. and Sarre, P. (1995). 'Black middle-class formation in contemporary Britain', in T. Butler and M. Savage (eds), *Social Change and the Middle Classes*, UCL Press, London, pp. 76-94.

Phizacklea, A. and Ram, M. (1996), 'Being your own boss: ethnic minority entrepreneurs in comparative perspective', *Work, Employment and Society*, 10 (2), 319-339.

Piore, M. and Sabel, C. (1984), *The Second Industrial Divide: Possibilities for Prosperity*, Basic Books, New York.

Putnam, D. (1993), *Making Democracy Work: Civic Traditions in Modern Italy*, Princeton University Press, Princeton, NJ.

Raghuram, P. and Hardill, I. (1998), 'Negotiating a business: case study of an Asian woman in the Midlands', *Women's Studies International Forum*, vol. 21, no. 5, pp. 475-483.

Raghuram, P. and Strange, A. (2000), *Studying Economic Institutions, Placing Cultural Politics: Methodological Issues from an East Midlands Study*, Paper presented to the RGS/IBG Annual Conference, University of Sussex.

Ram, M., Ford, M. and Hillin, G. (1997) 'Ethnic minority business development: a case from the inner city', in D. Deakins, P. Jennings and C. Mason (eds), *Small Firms: Entrepreneurship in the Nineties*, Paul Chapman, London, pp. 89-100.

Ram, M. and Jones, T. (1998), *Ethnic Minorities in Business*, Small Business Research Trust, London.

Razin, E. and Light, I. (1998), 'The income consequences of ethnic entrepreneurial concentrations', *Urban Geography*, vol. 8, pp. 554-576.

Salais, R. and Storper, M. (1993), *Les Mondes de Production*, Editions de l'Ecole des Hautes Etudes en Sciences Sociale, Paris.

Storey, D.J. (1994), *Understanding the Small Business Sector*, Routledge, London.

Tarlo, E. (1996), *Clothing Matters: Dress and Identity in India*, Hurst, London.

Chapter 13

Over- and Under-Embeddedness: Failures in Developing Mixed Embeddedness Among Israeli Arab Entrepreneurs

Michael Sofer and Izhak Schnell

Introduction

Arab industry in Israel has grown intensively during the 1970s and 1980s, more than doubling the number of operating enterprises, their average size and the form of production. In the same period, Arab industry experienced a restructuring process by which entrepreneurs shifted from mimicking strategies of operation to more competitive strategies based on efforts to exploit any slight opportunity in the market (Schnell *et al.* 1995). As a consequence, 71% of Israeli Arab entrepreneurs managed to break ethnic barriers and to access Jewish markets, and close to half of them even succeeded in expanding into markets in the national core (Schnell *et al.* 1999). Nonetheless, Arab entrepreneurs were rarely able to translate these achievements into a dynamic capital accumulation process based on improved multipliers of economic growth. This discrepancy calls for a better explanation of the barriers that Arab entrepreneurs – as an example of a marginal ethnic minority – are forced to overcome in their struggle for further integration into the state economy.

During the last decade an increasing number of studies confirm the hypothesis that inter-firm linkages and networks are key factors for business growth. Appropriate intensive linkages may generate, along other economic benefits, economic multipliers to the firm resulting in improved welfare levels to the local and the regional economy (Scott 1991, Felsenstein 1992, Staber and Schaefer 1996). A key issue in network studies is the degree to which firms are embedded in wider socially and politically structured milieus. In this context, particular attention should be devoted to minorities' need to insert themselves in two different and separate milieus. This is important owing to the key role that entrepreneurship plays in minorities' socio-economic mobility. Nothwithstanding the challenges and barriers that each milieu may offer, entrepreneurs' success in being embedded in both milieus and developing wide business linkages and networks is crucial for their integration into the economy as well as society (Aldrich and Waldinger 1990, Barrett *et al.* 1996). This form of mix-embeddedness is a target to be reached by ethnic entrepreneurs, who alternatively may find themselves trapped within ethnic enclaves, not being able to overcome barriers that stem from their ethnic origin and/or location in the national, regional and urban periphery (Kloosterman *et al.* 1999).

An example of business operating in two different milieus is the case of Arab entrepreneurs in Israel. They face both the wider Israeli capitalist economy, largely dominated by the Jewish sector, and their own socio-economic system. They have to face each sector's system of values and norms, which underpin their business cultures and power relations – the power of firms to determine exchange relations.

The aim of this chapter is to offer a conceptual framework for the analysis of the degree and form of embeddedness in the state economy of marginal minority group. Based on the experience of Israeli Arab entrepreneurs, we analyse some of the difficulties that minority entrepreneurs encounter in their attempts to break barriers of marginality and ethnicity and to participate in both minority and majority business cultures, politics and information networks. We offer the concept of 'over-embeddedness' to characterise those entrepreneurs whose commitments to kinship groups prevent them from exploiting opportunities in inter-ethnic markets. We also offer the concept of 'under-embeddedness' to characterise those entrepreneurs who fail to translate their wide and complex networks into an economic advantage in the markets, owing to lack of understanding of the Jewish business culture or absence of capability to improve their positions in economic exchange networks.

The data presented in this chapter are based on intensive field study conducted in the early 1990s and partly updated at the late 1990s. During the study, 514 entrepreneurs and plant managers, who made just under 60% of all Arab owned enterprises in Arab settlements at that time, were interviewed. An enterprise was defined as a production unit with at least three workers. The comprehensive questionnaire used to collect data included items relating to all components of production, sources of labour and capital, marketing and purchasing of inputs. Additional sources of data were open-ended interviews with 80 Arab industrial entrepreneurs representing different generations, branches and factory sizes. From these interviews information was gathered on strategies for survival, opinions on barriers to growth, and views on the futures of their enterprises.

Our argument is developed in five stages, beginning with a discussion of the theoretical framework for the analysis of mixed embeddedness within the framework of ethnic relations in Section 1, followed by a description of Arab industry in Israel in Section 2. Section 3 focuses on the evaluation of the structure and features of Arab enterprises' embeddedness in the Israeli economy, followed by a demonstration of situations of over- and under-embeddedness in Sections 4 and 5 respectively.

Embeddedness

Our argument is grounded in the contention that firm growth, and the resulting regional growth, requires firms to be embedded in concrete ongoing systems of social relations (Granovetter 1985). Several studies have provided evidence of the impact of social structure and culture on economic action. In these studies, the intensity and the dynamic of inter-firm linkages and networks are treated as indicators for economic growth (Scott 1991, Grabher 1993, Taylor 1995, Oinas 1999), while external networks have a particularly decisive impact on small plants' growth chances (Kay 1993, Hardill *et al.* 1995). The key issue in these investigations is the degree to which firms are embedded in various markets through their relationships with competitors, suppliers, other business organisations and public decision-making forums, as well as with members of their communities (Best 1990, Harrison 1992, Markusen 1994, Lakshmanan and Okumura 1995).

Embeddedness tends to refer to the cultural, socially structured and institutional milieus in which entrepreneurs perform as economic agents (Todtling 1994). From the cultural perspective, embeddedness may be viewed as processes in which agents acquire customs, habits, or norms in an unerring way that unintentionally determine their decisions and behaviours, and structure awareness of their relevant milieus. From the structural perspective, embedded networks may be characterised by agents' connectedness, reciprocity, interdependence, autonomy, and power relations in terms of control over both economic and social relations (Grabher 1993, Portes and Sensenbrenner 1993, Oinas 1999). From the institutional perspective, embeddedness relates to agents' accessibility to education and training institutions, incubation and innovation centres, market organisations, business associations, and business practices which regulate particular markets (Todtling 1994, Kloosterman *et al*. 1999).

When considering networking of small firms, attention has been given to economic and non-economic linkages within the wider milieu (Curran and Blackburn 1994). These linkages are of two main types. First, those related to the firm themselves, such as membership of local chambers of commerce, trade associations, and so on. Second, those not related to the business directly, but which reflect other embedded relations such as owners' social relations, political party membership, leisure activities and friendship and family relations that are still relevant to economic activities. Business owners are 'networkers' who participate in and sustain a web of economic and social links for information, customers and suppliers. The strength of these linkages reflects their level of integration into and embeddedness within an economy. The sum of their connections is interpreted as underpinning a vibrant local economy with the characteristics of an 'industrial district' (Curran and Blackburn 1994).

In these terms, the socio-cultural 'mixed embeddedness' of ethnic entrepreneurs is very different because they are forced to operate in more than one milieu, each of which is structured by different sets of meanings, power relations and with different degrees of openness for inter-ethnic or trans-ethnic encounters. Aldrich and Waldinger (1995) offer a theoretical framework to understand the structural processes shaping ethnic economies and entrepreneurship in which they stress three major processes. First, they identify the 'opportunity structure' formed in an historical context and by political decisions that may create different forms of capitalist accumulation on the two sides of an ethnic divide. Second, they recognise that access to firm ownership may channel ethnic groups either to low-reward enterprises or to a segregated ethnic economy, depending upon the degree of inter-ethnic competition. Third, they recognise that ethnic group characteristics may steer a group's orientation toward entrepreneurship.

Ethnic entrepreneurs frequently tend to concentrate their economic activities within their ethnic milieu as it is implied by the ethnic enclave hypothesis, but even in these cases they are not able to escape the majority institutional milieu of the city and/or the state in which they operate. When entrepreneurs try to escape their limited ethnic enclave economy, they are forced to keep their links with their ethnic milieu and at the same time to establish links with the majority milieu. Only by understanding entrepreneurs' need to operate within and outside their own milieu may embeddedness encompass its original meaning concerning the interplay among economic, social and institutional contexts of ethnic entrepreneurship (Kloosterman *et al.* 1999).

Talmud and Yanovitzky (1998) offer a more specific model for the analysis of firm's embeddedness. In line with Grabher (1993) and Oinas (1999) they analyse networks within cultural and power fields. Networks, according to them, should be analysed less in terms of numbers of links and intersections and more in terms of power relations and horizons of the ·players' awareness. Moreover, embeddedness is related to different sectors of society: political and economic elites and institutions, structures of market networks and so on. Politically, evidence shows that firms closely attached to political centres succeed better in appropriating benefits for their operations and in developing greater autonomy in their relations with governments (Han 1992, Talmud and Mesch 1997). Market autonomy may be achieved by maximising the flow of information from a variety of sources about market opportunities and conditions, as well as by developing reciprocal relations rather than relations of dependency (Burt 1992).

Applying the embeddedness ideas to the case of the Israeli Arab economy requires recognition of the deep cultural, economic and political divide between the Jewish and the Arab milieus. Arabs are economically, politically and culturally marginalised (Gradus *et al.* 1993, Falah 1993, Haidar 1993) in an economy that is highly politicised (Aharoni 1991). Entrepreneurs, particularly those who have escaped the mimicking strategy, act like the goddess Yanus, turning their faces toward the two different milieus (Figure 13.1). This form of mixed embeddedness means that in each milieu they are required to link themselves to the economic and political elites and institutions, as well as to the players in the markets. Embeddedness within such milieus, as shown in Figure 13.1, may be measured in terms of the complexity and intensity of networks, power relations and horizons of awareness concerning the codes of the two business cultures, opportunities and risks. It is basic to our argument that well-balanced mixed embeddedness means mutual and integrated co-ordination of the three dimensions of *networks, power and horizons of awareness*. Lagging behind on any one of them may lead to an imbalance resulting in over- or under-embeddedness.

A major issue concerns the level of embeddedness required by any firm in order to survive (Grabher 1993). Too little embeddedness may expose networks to the erosion of their supportive tissue of social practices and institutions. Too much embeddedness, however, may promote a petrifaction of this supportive tissue and, hence, may convert networks into cohesive coalitions against more radical innovation and even major change.

Here, therefore, we offer two concepts that can be used to characterise imbalances in entrepreneurs' embeddedness. First, we define *over-embeddedness* as a situation in which entrepreneurs manoeuvre their kinship and community systems to support their entrepreneurship, but the resulting commitment impedes their participation in inter-ethnic markets. Second, we define *under-embeddedness* as entrepreneurs' success in developing intensive and complex inter-ethnic networks while failing to adequately evaluate business opportunities and/or to gain enough power to translate their networks' relationships into economic growth.

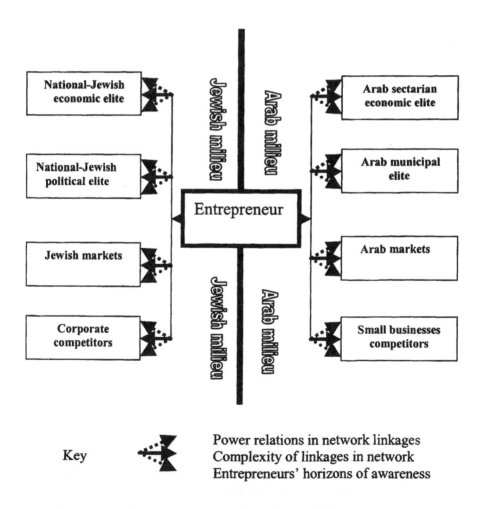

Figure 13. 1 A general model of mixed embeddedness

Arab Industry in Israel

During the last five decades Arab industry in Israel has gone through a sequence of three restructuring stages influenced by the development of the larger Israeli economy. Three major mechanisms influenced these restructuring phases: majority-minority relations, core-periphery relations, and selective government policies (Sofer and Schnell 2000). These three mechanisms overlap and they have combined to shape the form of Arab industrialisation, including the branches of activity that are selected, plant formation and entrepreneurial style.

With the establishment of the state of Israel, the Arab, mainly rural, population that remained was forced to return to traditional peasant economy supplemented by small local industry including coal and lime production, stone quarrying, oil pressing, flour milling and pottery (Zarhi and Achiezra 1966). The Arab economy was completely neglected and martial law blocked any opportunity for the development of external economic networks. This created the preconditions for reintegrating this sector as a subordinated periphery in later years. Exclusion turned Arab industry into an ethnic enclave economy.

During the first restructuring stage, which began in the late 1950s, the state managed capitalist system split the national periphery into Jewish and Arab segments. Large industrial and infrastructure projects were initiated in the Jewish periphery with direct government investment, while in the Arab settlements, the government avoided any large investment. Faced with limited opportunities, Arab labour commuted to the subordinate segments of the labour markets of the Jewish core.

During the second stage in the 1970s, the monopoly power of emerging large industrial corporations affected the spatial pattern of industrial production. Arab settlements did not gain from the shift of industrial enterprises to the periphery because of a scarcity of available land for industry (Sofer *et al.* 1996), and an absence of national subsidies (Bar-El 1993). However, they did gain sewing shops that acted as subcontractors to textile corporations. A strategy of growth from below was adopted (Khamaisi 1984, Falah 1993), that led to the rapid growth in numbers of new plants in the food and construction sectors (Table 13.1) coupled with the adoption of new styles of entrepreneurship that involved innovation and risk-taking. Most plants doubled their size to an average of 15 workers. Women joined the labour force. Managers began to use more formal management and marketing techniques, and divided production into specialised lines. Businesses were confined to traditional sectors, but Arab entrepreneurs proved their ability to compete on Jewish markets and increased their penetration of those markets including those of metropolitan areas (Schnell *et al.* 1999).

Table 13.1 The distribution of Arab-owned industrial plants by main branches, 1983-1992

	Percent of plants		
Branch	1983	1990	1992
Food & beverages	18	19	19
Construction materials	31	37	34
Metals & iron	11	10	8
Textiles & clothing	36	27	26
Others	4	7	13
Total	100	100	100
No. of plants	415	829	900

Sources: Meyer-Brodnitz and Czamanski (1986), Atrash (1992), Schnell *et al.* (1995).

During the 1990s, the Israeli economy has hesitantly started to experience a third phase of restructuring associated with globalisation. A shift of industrial interests towards Middle Eastern labour markets, brought the establishment of some 30 projects (especially sewing shops) in Jordan, Egypt and the Palestinian territories. Most of these projects resulted from the relocation of standardised production processes formerly based in Arab settlements. This restructuring phase has had a destructive impact on Arab entrepreneurs in Israel owing to their non-participation in (though some may say exclusion from) the new opportunities offered by the peace process and globalisation. Cautious estimates suggest that between 50% and 60% of the medium- and large-sized textile plants and sweatshops in Arab settlements have been closed since 1995, and a much larger proportion of the labour force employed in this industry has been laid off. Despite the visible signs of its growth potential in the early 1960s, Israeli Arab industry did not show any clear signs of shifting towards hi-tech or sophisticated industrial branches in the last decade of the

20ᵗʰ century. Arab industry persists in the same sectors, a response to the changing economic milieu from their marginal position.

Industry, then, is spread among the 60 settlements that represent the majority of Arab settlements with populations over 5,000. Between 1983 and 1992, plants employing three or more workers doubled in number, reaching more than 900 (Table 13.1), with the size of the industrial workforce employed in Arab settlements growing to about 12,000 by 1990 (Atrash 1992). National figures indicate that in the 1990s, Arab-owned enterprises represented nearly 6% of all Israeli plants employing five or more workers (Central Bureau of Statistics 1992).

Table 13.2 Sector specialisation index of the labour force in Arab-owned plants with three or more employees, 1992*

Specialised Sectors		Marginal Sectors	
Textiles & clothing	4.2	Food & beverages	0.4
Construction materials	3.7	Metal products	0.2
Woodworking	1.5	Paper & printing	0.2
		Rubber & plastics	0.2
		Electronics & electricity	0.002

*   The figures for Israeli industry are from 1989, those for Arab industry from 1992. The Specialisation Index is the rate of employee distribution by branch in the Arab sector, compared with the branch distribution in Israeli industry as a whole.

Source: Schnell *et al.* (1995).

Typically, small household production and informal subcontracting activities, which constitute 43% of enterprises in this sector, exist side by side with an increasing number of workshops and factories (around 57%). In the late 1980s, a very small number of large-scale enterprises have emerged (about 1% of the total which employ more than 50 workers each). The pattern of sectoral specialisation is shown in Table 13.2.

A significant proportion of plants (60 to 80% in most sectors) are located in residential areas, especially on the ground floor of the owners' homes or in rented residential buildings. Where an industrial area is available, it is of inferior quality, lacking appropriate physical infrastructure (Sofer *et al.* 1996). In 1992, the average number of employees per plant was about 15, a high proportion of whom were women, employed primarily in sewing shops. Despite the increase in their mobility, most women work in the settlement where they live (Atrash 1992). The formal training of the labour force is limited, with less than 2% having any formal academic or technical education. Most of the training of the professional workers is done on the shop floor. The most common type of ownership is individual or family (83%), where several brothers own and manage a plant.

Arab entrepreneurs do not usually obtain investment capital from financial institutions. Personal savings are the most common source of initial capital investment, while other family members are also an important source. Only 14% percent sought bank loans as a major source of investment. Profits are the most common source of capital for the further expansion of production (Schnell *et al.* 1995).

The Networks of Arab Enterprises

In evaluating the degree of embeddedness of Arab enterprises we concentrate in this section only on the analysis of the structure of their *diversified business networks* – the complexity of their linkages and networks (as shown in Figure 13.1). The analysis of the other two dimensions of embeddedness – *power* and *awareness* – is undertaken in subsequent sections based on Arab entrepreneurs' experience, and deal with the conditions of over- and under-embeddedness that Arab entrepreneurs are faced with.

Empirical evidence sheds light on the intensity and structure of networks related to business information, purchasing and sale linkages, and labour force recruitment. It shows that information on technological innovations is provided by suppliers to about 60% of the entrepreneurs, and about 28% of them gather such information from other sources. These sources include professional journals (27%), other publications and television programmes (9%), but only 7% from competitors – mainly small firms in the same settlement. Other sources of information remain marginal (Abo Sharkia 1996). Suppliers have an interest in manipulating their potential clients, and Arab entrepreneurs have no reasonably unbiased evaluation capability.

Local networks play a major part in the recruitment of employees. The home settlement provides three-quarters of the labour force, with one third of them coming from the employer's clan (*hamula*). Employees from the extended family, and in some cases from the *hamula*, serve as managers and clerks, while the majority of manual workers are from other local families. The latter are employed in under-privileged jobs that pay low salaries, frequently below the level of the minimum wage owing to the low profitability of most Arab enterprises. Consequently, Arab skilled labourers prefer to commute to the Jewish hubs of industry where better opportunities are offered. Jewish employees are less than 1% of the labour force in Arab enterprises, mainly in specific jobs such as engineers and sales managers, especially when the Jewish market is targeted.

In analysing Arab entrepreneurs' purchasing and sale links we have subdivided the market into meaningful sub-markets based on major structural barriers within the Israeli space economy relating to ethnicity, marginality and regional scale (Schnell *et al.* 1999). Accordingly, Arab entrepreneurs can be interpreted as having developed nine sub-markets, though in practice they have expanded into only five of them (Table 13.3).

Table 13. 3 The purchasing and sale links of Arab entrepreneurs with different sub-markets

Sub-markets	% Purchase Links	% Sales Links
Intra-settlement	2	22
Arab home region	15	39
Neighbouring Jewish region	25	18
National core	45	15
Arab inter-regional	6	3
Jewish inter-regional	3	1
Palestinian markets	0	1
Jews in occupied territories	0	0
Overseas	4	1

The data demonstrate Arab industry's high dependency on inputs supplied by national corporations based in the national core, and on subcontracted production lines from new towns in the Jewish periphery. Textiles, cement and other raw materials for the construction industry are purchased from external suppliers. The Arab economy supplies less than one quarter of the raw materials, leaving most of the profits in the Jewish sector. For sales, Arab entrepreneurs manage to expand into more diverse

markets. Although close to two thirds of their sales are directed to the Arab
peripheral regions, they have expanded more into Jewish markets, than into
either more distant Arab markets within Israel or into Palestinian markets.
Most enterprises tend to sell to two or three markets, with 13% selling to
four and five markets (Table 13.4). All the textile enterprises, and close to
one third of all other businesses which sell only to one market, have chosen
the more rewarding but risky Jewish market. And, more than one third of
those that sell to two markets sell to at least one Jewish market. Almost all
enterprises that sell to more than two markets sell also to Jewish markets.
This means that ethnic barriers are frequently overcome as many of the
entrepreneurs confirmed at interview. They stressed that they must first
prove themselves, but once they start to win markets discriminatory
attitudes based on ethnicity become irrelevant.

**Table 13.4 Enterprises by number of markets and percent overcoming
barriers**

No. of markets	% plants	% overcoming barrier of:	
		Ethnicity	Peripherality
1	17	33 (98)*	0 (93)*
2	35	34	24
3	35	92	48
4+	13	100	48
All markets	100	71	48

*The first number is for non-textile enterprises and the second for textile enterprises.

A major conclusion is, therefore, that Arab entrepreneurs in Israel
have developed wide and complex sale linkages compared to the relatively
limited networks in other aspects of the entrepreneurial process. A
question to be asked here is whether they have achieved reciprocal power
relations in the business milieus they operate in, and whether they have
succeeded in breaking barriers to information flows, evaluation capabilities
and other aspects of those two business cultures. To explore these issues the
analysis shifts to information from interviews with Arab entrepreneurs and
questions of over- and under-embeddedness.

Over-Embedded Entrepreneurs in the Arab Milieu

Over-embeddedness within the Arab milieu is a condition that many Arab entrepreneurs have to confront. Since most of them opened their businesses with intensive support of the extended family and their home community, they feel indebted to their supporters once they have started to succeed. The story of one entrepreneur demonstrates the dilemma:

> I've been in the field since 1975. I started working at the age of 16 in my hometown. After a while I moved to work for Gibor (a Jewish textile corporation). Once I had learned the job I moved to work as a seamstress in a workshop in my town. It took me only a short time to become in charge of 40 seamstresses. I saved any possible Shekel to buy a sewing machine to do some extra work. During the next decade I bought two more sewing machines with the help of my sister and I asked my sister and another woman from my family to help me. The three of us sewed clothes in my bedroom for a contractor from Haifa. My sister's husband joined us. He searched for clients. At the beginning it was difficult till we got from one company 200 pieces which we sewed to their satisfaction. Gaining their trust opened for us the route to success. My younger brother joined us, and then my cousin, who did all the bookkeeping. I owe my family a lot. For a decade we did not earn much. All the money was reinvested in buying sewing machines. We used the family bedroom and a closed veranda. The only thing I helped my family with was the payments for my brother's education. On the other hand, another brother gave me 10,000 Shekels to buy a car for the business.

This example describes the situation of many Arab entrepreneurs. The project was made possible by the intensive support given by the extended family. They provided money, work and personal sacrifice. As a consequence, the entrepreneurs are obliged to support their relatives, and they cannot leave them aside once they succeed. Other typical forms of support are, providing space (usually at the ground floor of the extended family house), paying for the entrepreneur's everyday living expenses, and moral support. In most cases, wider kinship units also provide support. For example, entrepreneurs can rely on the clan as a guaranteed market. In some cases, they may obtain release from paying municipal taxes, particularly when the mayor belongs to their *hamula*, and to operate a business from the family house without a permit.

The absence of risk-reducing mechanisms makes the extended family the main source of entrepreneurial support but, at the same time, the family imposes demands on successful entrepreneurs that prevent them from achieving further economic growth. Two examples illustrate the importance of family commitments:

> My brother helped me to get through school. I graduated as a practical
> engineer in electronics, but there's no work for Arabs in the field. So I was
> forced to start a business. I did it also to help my two brothers. One didn't
> pass his Law license tests and one didn't acquire any profession. They helped
> me in saving money for the business, particularly the one who quit school and
> went to work, so now I feel obliged to help them.

This example emphasises the mutual support of brothers to develop
economic opportunities for the extended family. The second example
emphasises the role of the extended family in managing the enterprise:

> I am the eldest. Two brothers work with me, and three sisters sew. I am the
> manager but much of my time I search for customers. One brother manages
> the production line and the other the books. My sister learns now to use the
> computer. Each of us can cover the responsibility of the other brothers, a
> situation that gives us a lot of flexibility. As you see the factory is a family
> project and its success is the success of the whole family.

But the involvement of family relations in the decision-making of the
enterprises may cause some problems. A worker in one of the more
successful firms in the Arab sector complained:

> There are two many managers. So many family members are involved, with
> each of them giving you orders. You do not know whom to obey and who has
> the real authority. They take high salaries but they barely contribute to the
> firm. Nobody can complain because they are all one family.

The same complaints can also be heard from an owner of an
enterprise:

> It turned out to be a big obstacle to work with my brothers. I have to consult
> them, to accept ideas that I know are harmful for the business, and I cannot
> fire them. I owe them so much and any problem at work becomes
> immediately a family issue. From time to time one of my brothers needs some
> money urgently for a wedding, school or any other purpose, and it forces us to
> hold further development. I feel that the partnership with my brothers enabled
> the establishment of the enterprise in the beginning, but it prohibits further
> development today when we are in a new era in which each family looks for
> itself.

The 'warm hug' of the Arab milieu is also evident at the community
level. The majority of enterprises are located on the ground floor of

extended families' houses with, the family being patient with disturbances and the Municipality turning a blind eye to the breaking of regulations:

> My father, who lives in the first floor of the house, agreed to open a sewing plant on the ground floor, and so did my other two brothers. I closed the floor and started to work. Last year I added 200 square meters to the plant and no body bothered to complain. ... We are the largest *hamula* in the town and we have controlled the Municipality for the last decade. We developed infrastructure so people are satisfied. So, I did not need a permit for the construction. Concerning municipal taxes we compromise. (Here he made a meaningful sign I had to interpret on my own.)

Close communal ties may work even further in favour of the entrepreneur:

> My workers will not complain even when business is bad and I have to pay them less money. They are from the family and they would be ashamed to complain about a family member. This will put shame on the whole family. Customers will be embarrassed to go and buy building blocks elsewhere. They know it will become a family matter.

But, the 'warm hug' of the community may prohibit further development. An entrepreneur who is located in a residential area and who wants to expand his business frequently finds it impossible. Roads are too narrow, infrastructure is inadequate and trucks cannot reach most houses. He is forced to move out to an industrial zone, which rarely exist in the Arab towns. The dilemma for an entrepreneur who tried to expand into Jewish markets demonstrates the problem associated with this over-embeddedness:

> In order to expand I had to move to an industrial area. My place was too small for the plant. We have an industrial area, but with no infrastructure. There is no sewerage, roads, and electricity, not mentioning a building. To build all of it on my own is too expensive for me. So, I decided to check out the possibility to shift to the industrial zone of the neighbouring Jewish town. The extra expenses, for renting the place, the permit etc., were so high that paying them would have forced me to sell at higher prices. This experience made me afraid that another small enterprise could be established in an entrepreneur's family home in our town, stealing from me my local clients, before I would gain even one new client outside my home town.

Another aspect of over-embeddedness depends on entrepreneurs' horizons of awareness, as these are constituted in the situation of a mixed milieu. An older entrepreneur told us:

> Jews refuse to buy concrete blocks from me because I am an Arab. They discriminate against me because they prefer to provide work for Jews. My blocks are significantly cheaper and their quality is higher than the regular standard. Still I don't get any Jewish customers. (Question: We do not see any quality control signature on your product, don't you think that this is necessary in order to attract Jewish customers?) Answer: Everybody in the town knows me and knows the quality of my product. I don't have to pay 10,000 Shekels for the quality control approval seal and Jews will never trust me anyway because I am an Arab.

The entrepreneur was unable to understand the differences between Arab and Jewish business cultures. For him, trust based on his kinship origin should have been enough, while formal institutions of quality control were not perceived to be relevant for his reputation. For most Jewish customers this is the only guarantee of the producer's faith in the quality of his products.

Arab entrepreneurs gain kudos in the community at large as an economic elite. This captures the imagination of many frustrated youngsters who view entrepreneurship as a major route for mobility within Israeli society. Even Moslem fundamentalists develop symbiotic relations with entrepreneurs by asking them to donate to Mosques and community services. Entrepreneurs succeeded in turning this support into an economic advantage, gaining privileged standing within the family and the local political elite. Over-embeddedness then is articulated in:

- the dominance of intra-ethnic patterning of business networks;
- the power that entrepreneurs gain in their intra-ethnic milieu compared to their weakness in inter-ethnic milieus (as will be demonstrated in the next section); and
- barriers imposed on entrepreneurs' horizons of awareness.

The advantages that Arab entrepreneurs gain in their intra-ethnic milieu encourage them to play according the informal norms of conduct required in this specific reality, and to enjoy independence even when it involves the low profitability typical to local markets. The persistence of this pattern of relationships reinforces the ethnic (separatist) business culture which, in turn, reduces entrepreneurs' chances to compete in the more rewarding markets of the Jewish milieu, where they must play by different rules. Many entrepreneurs refuse to take the risks and retain only intra-ethnic networks. Others, mainly younger generation entrepreneurs, are highly motivated to make any effort and to make personal sacrifices and take risks in breaking ethnic barriers (Schnell *et al.* 1995). In other words, the over-embedded entrepreneur tends to remain contained in the safe and supporting milieu of his intra-ethnic stronghold. The over-embeddedness of Arab entrepreneurs may be initiated voluntarily for specific reasons such as the political motivation to develop an autonomous economy. However, over-embeddedness can be seen as a survival strategy, adopted when barriers of ethnicity and peripherality are seen as being unbreakable.

To sum up, the small local markets and intense competition force Arab entrepreneurs to cut costs, exploit kinship ties, and pay reduced wages, a situation typical of informal economic activities. The persistence of such a strategy may strengthen traditional institutions, but it may also widen the gap, in terms of economic behaviour, between the Arab and Jewish milieus, with the consequence of reducing the prospects for integration across the inter-ethnic market.

Under-Embeddedness in the Jewish Milieu

Arab entrepreneurs' experiences of under-embeddedness surface almost entirely in the context of inter-ethnic networks. As we have shown, close to three-quarters of Arab enterprises sell their output to Jewish markets and almost half sell to the national core. About half of the enterprises sell to three or more different sub-markets. These results suggest that Arab enterprises have developed intensive and complex networks with various segments of Israel's national economy, including the relatively more privileged sectors. Nevertheless, Arab entrepreneurs fail to overcome their marginal position. The question is, therefore, what can explain the apparent contradiction between Arab entrepreneurs' membership of relatively complex networks, including inter-ethnic links, and their lack of economic growth? Our argument is that this failure is the result of the subordinate position Arab entrepreneurs occupy in Jewish markets. They are cut off from information flows and sources to evaluate opportunities, and they have little power within business networks. This is the setting for the creation of under-embedded entrepreneurship. In this section we demonstrate the powerlessness associated with under-embeddedness that includes the inability of entrepreneurs to evaluate economic situations beyond their local horizons.

Elsewhere we have identified three strategies used by Arab entrepreneurs to deal with cultural differences (Schnell *et al.* 1995). A first strategy, used by food enterprises, is to produce Arab traditional products by using modern machinery. Traditional know-how, learned from grandmothers, is transformed into a modern production system. Such entrepreneurs base themselves in local networks and markets, but they also search inter-ethnic networks without taking great risks. As one respondent explained:

> The idea to open a cheese factory crossed my mind when my father told me how grandma had made great cheese from sheep's milk. I used her recipe and I started to sell to people in the town. After a while, I bought machines and I bought milk from other farmers in the village in order to increase production. We sell in most of the Arab settlements in the region and we even have Jewish customers. They come on Saturdays to buy in the shops in our town and we sell to them directly from the plant. Once a week I go to Tel Aviv to buy some raw materials, so I sell cheese to some stores at the market. On my way home I also sell to a store in Nataniya.

A second strategy is exemplified by a number of construction enterprises that were able to take advantage of changes in the Jewish market. The government promoted a self-construction housing policy, in the hope of encouraging people in 'new towns' and the rural population to build new private houses and remain in the peripheral regions of the country. This policy followed a sub-urbanisation trend that started in the 1980s. Builders were ready to hire Arab construction contractors and to purchase construction material from Arab plants in their regions. In addition, the large expansion of the construction industry as a result of the massive immigration from the former Soviet Union created shortages in construction materials. These events opened a niche for Arab construction plants in the Jewish sector, which they have efficiently exploited until the economic recession of the mid-1990s. The following example illustrates the strategy:

> Muhammad's factory is a good example. As a small firm they realised the new opportunities. They hired a Jewish economist who helped them to develop a business strategy and to find clients in the Jewish settlements in the area. We have to admit we do not have enough financing and marketing expertise. Some of us used kinship contacts with Arab construction contractors who built in Jewish settlements, but this is not the same. Today they sell for millions of dollars. They modernised their machines and Muhammad is a real rich person.

The construction industry example also emphasises the fact that Jewish markets are not necessarily accessible. Furthermore, this example also shows the gap between business cultures that narrows Arab entrepreneurs' horizons of awareness of the opportunities in the Jewish milieu.

The third strategy is that of entrepreneurs in the textiles industry who operate as subcontractors to Jewish corporations. Such enterprises, however, have problems they must confront, as the following example demonstrates:

> When our plant gained the trust of our mother corporation we received a loan from them to buy sewing machines and we grew fast. I sent one brother to study machine engineering, my other brother to study management. The rest of the family accumulated 50,000 Shekels to help me. We received the support of workers from my extended family that manages women from the occupied territories. We are highly reliable producers and we gained very good reputation everywhere. Now I plan to produce on my own and to sell to stores. I have to do it because I have only extremely marginal profits as a subcontractor. I need more income to be able to help back my family who have helped me.

So far no entrepreneur operating in the textiles and clothing sector has been able to make it as an independent producer. These entrepreneurs have faced the dilemma of focusing their efforts. Locating in Tel Aviv, searching for customers, attempting to follow leading enterprises and new trends in fashion, means losing control over their production lines and even facing sanctions from their mother corporations. Yet, they remain in their home settlements, they continue to be cut off from information that could assist in the evaluation of financial and market possibilities. This dilemma exposes the difficulties in expanding Arab entrepreneurs' horizons of awareness into the Jewish milieu, which makes it hard for them to develop evaluation capabilities. In addition, it emphasises their dependency on mother corporations and/or customers in the Jewish sector, and their lack of power to create for themselves a degree of autonomy in the market. Competition with corporations has been identified as the hardest obstacle for businesses seeking to embed themselves in the inter-ethnic milieu. This difficulty is well demonstrated by a bakery owner that tried to overcome this barrier:

> In an attempt to grow I started to convince Jewish food stores in the region to buy Arab bread (pittas) from me. I promised storeowners to bring them fresh pittas twice a day, and I made them a little bit cheaper. For a while my business grew rapidly and it was a great success. I even started to plan the expansion of the bakery in an industrial zone, but then the problems started.

The bread bakery, which has a monopoly on standard bread, put pressure on the stores not to buy pittas from me. To make the story short, I was forced to compromise with them and to distribute pittas through them, with them sharing part of my profit.

The interviews with the sample group suggests that the lack of developed industrial areas, investment capital, and government subsidies equivalent to those given to neighbouring development towns, are the three major barriers to Arab enterprise development (Schnell *et al.* 1995). But, the in-depth interviews shed a different light on Arab entrepreneurs' failure to embed themselves in the inter-ethnic milieu. While there are over-embedded entrepreneurs who avoid any attempt to break away from the safe intra-ethnic market, others are willing to make any necessary effort to break these barriers. They even succeed in exploiting marginal niches in Jewish markets, but they fail to expand into the more rewarding markets by virtue of their under-embeddedness. Failure to determine sale conditions, dependency on mother plants and other reasons, leave them with low profitability and a low rate of growth. As a consequence, they cannot afford to take bank loans, and to pay the necessary expenses to operate as an officially registered enterprise in developed industrial zones. It is in this context that interviewee's comments should be interpreted.

Conclusions

In the mixed embeddedness model we present here we suggest that ethnic entrepreneurs' networks should be viewed on three complementary dimensions: (1) the intensity and complexity of networks; (2) power relations; and (3) entrepreneurs' horizons of awareness. This study shows that inter-ethnic networks do not give ethnic entrepreneurs instantaneous advantages unless they learn to operate outside their business culture, and unless they develop some business autonomy that may enable them to improve their bargaining position in the market. The fact that Arab entrepreneurs are rarely members of business organisations, like the Industrialists' Association, the Bureau of Commerce and other national small business organisations, does not assist them to secure better access to the national economic and political elite. Arab entrepreneurs only rarely make use of support programmes offered by the Ministry of Industry and Commerce because they do not believe they have the ability to meet their prerequisites. These features emphasise the absence of links with the Jewish and the national economy and the country's political elite, as well as with Jewish competitors. In this context, it is understandable that Arab entrepreneurs fail to translate into profits their efforts and ability to expand into Jewish markets.

It could be said that Arab entrepreneurs might improve their awareness of opportunities in Jewish markets if they adapted to Jewish business culture. However, they have little chance of improving their power relations with national corporations without public support. Since the restructuring of an intra-ethnic informal business culture is a response to powerlessness in inter-ethnic networks, it seems that Arab entrepreneurs are caught in a vicious circle of marginalisation. Interviews with managers of Jewish corporations who tried to initiate partnerships with Arab entrepreneurs show that almost all of them have failed because of their inability to get used to informal forms of management. Therefore, there is an urgent need to initiate special programmes to integrate Arab enterprises into Israel's national economy – integration that might also serve as an appropriate mechanism for socio-political integration. The absence of such programmes may halt the proper and balanced integration of Arab entrepreneurs into the wider Israeli economic milieu, leaving them to retreat into entrenched over-embeddedness within the intra-ethnic Arab milieu.

References

Abo Sharkia, N. (1998), *Small Businesses and Their Networks among the Arab Minority in Israel*, unpublished Master Thesis, Department of Architecture and Town Planning, Technion, Haifa.

Aharoni, I. (1991), *The Political Economy in Israel*, Am Oved, Tel-Aviv (in Hebrew).

Aldrich, H.E. and Waldinger, R. (1990), 'Ethnicity and entrepreneurship', *Annual Review of Sociology*, vol. 16, pp. 111-135.

Atrash, A. (1992), 'The Arab industry in Israel: Branch structure, employment and plant formation', *Economics Quarterly*, vol. 152, pp. 112-120 (in Hebrew).

Bar-El, R. (1993), *Economic Development in the Arab Sector*, The Jewish-Arab Centre for Economic Development, Tel Aviv (in Hebrew).

Barrett, G.A., Jones, T.V. and McEvoy, D. (1996), 'Ethnic minority business: Theoretical discourse in Britain and North America', *Urban Studies*, vol. 33, no. 4-5, pp. 783-809.

Best, M. (1990), *The New Competition: Institutions of Industrial Restructuring*, Harvard University Press, Cambridge, MA.

Burt, R.S. (1992), *Structural Holes: The Social Structure of Competition*, Harvard University Press, Cambridge, MA.

Central Bureau of Statistics (1992), *Statistical Abstracts of Israel*, Jerusalem.

Curran, J. and Blackburn, R. (1994), *Small Firms and Local Economic Networks*, Paul Chapman, London.

Falah, G. (1993), 'Trends in the urbanization of Arab settlements in Galilee', *Urban Geography*, vol. 14, no. 2, pp. 145-164.

Felsenstein, D. (1992), 'Assessing the effectiveness of small business financing schemes: Some evidence from Israel, *Small Business Economics*, vol. 4, pp. 273-285.

Grabher, G. (1993), *The Embedded Firm*, London, Routledge.
Gradus, Y., Razin, E. and Krakover, S. (1993), *The Industrial Geography of Israel*, Routledge, London.
Granovetter, M. (1985), 'Economic and social structure: The problem of embeddedness', *American Journal of Sociology*, vol. 91, no. 3, pp. 481-510.
Haidar, A. (1993), *Obstacles to Economic Development in the Arab Sector in Israel*, The Israeli Arab Centre for Economic Development, Tel-Aviv (in Hebrew).
Han, S.K. (1992), 'Curning firms in stable markets', *Social Science Research*, vol. 21, pp. 406-418.
Hardill, I., Fletcher, D. and Montagne-Villette, S. (1995), '"Small firms", distinctive capabilities and the socioeconomic milieu: Findings from case studies in Le Choletais (France), and the East Midlands (UK)', *Entrepreneurship and Regional Development*, vol. 7, pp. 167-186.
Harrison, B. (1992), 'Industrial districts: old wine in a new bottle?', *Regional Studies*, vol. 26, pp. 469-483.
Kay, J. (1993), *Foundations of Corporate Success*, Oxford University Press, Oxford.
Khamaisi, R. (1984), *Arab Industry in Israel*, Unpublished Masters Thesis, Technion, Haifa (in Hebrew).
Kloosterman, R., van der Leun, J. and Rath, J. (1999), 'Mixed embeddedness: Informal economic activities and immigrant businesses in the Netherlands', *International Journal of Urban and Regional Research*, vol. 23, no. 2, pp. 252-266.
Lakshmanan, T.R. and Okumura, M. (1995), 'The nature and evolution of knowledge networks in Japanese manufacturing', *Papers in Regional Science*, vol. 74, pp. 63-86.
Markusen, A. (1994), 'Studying regions by studying firms', *Professional Geographer*, vol. 46, no. 4, pp. 477-490.
Meyer-Brodnitz, M.B. and Czamanski, D.T. (1986), *Economic Development in the Arab Sector in Israel*, Centre for Urban and Regional Studies, Technion, Haifa (in Hebrew).
Oinas, P. (1999), 'Voices and silences: the problem of access to embeddedness', *Geoforum*, vol. 30, pp. 351-361.
Portes, A. and Sensenbrenner, J. (1993), 'Embeddedness and immigration: Notes on the social determinants of economic action', *American Journal of Sociology*, vol. 98, no. 6, pp. 1320-1350.
Schnell, I., Benenson, I. and Sofer, M. (1999), 'The spatial pattern of Arab industrial markets in Israel', *Annals of the Association of American Geographers*, vol. 89, no. 2, pp. 311-336.
Schnell, I., Sofer, M. and Drori, I. (1995), *Arab Industrialization in Israel*, Praeger, Westport.
Scott, A.J. (1991), 'The aerospace-electronics industrial complex of Southern California: The formative years (1940-1960)', *Research Policy*, vol. 20, pp. 439-456.
Sofer, M., Schnell, I. and Drori, I. (1996), 'Industrial zones and Arab industrialization in Israel', *Human Organization*, vol. 55, no. 4, pp. 465-474.
Sofer, M. and Schnell, I. (2000), 'The Restructuring stages of Israeli Arab industrial entrepreneurship', *Environment and Planning A*, vol. 32, pp. 2231-2250.
Staber, U. and Schaefer, N. (eds), (1996), *Business Networks: Prospects for Regional Development*, Walter de Gruyer, Berlin.
Talmud, I. and Mesh, G.S. (1997), 'Market organization and corporate instability: the ecology of inter-industrial networks', *Social Science Research*, vol. 26, pp. 419-441.
Talmud, I. and Yanovitzki, I. (1998), 'The contradictory demand paradox: social embeddedness and organizational performances', *Israeli Sociology*, vol. 1, no.1, pp. 55-90 (in Hebrew).

Taylor, M. (1995), 'The business enterprise, power and patterns of geographical industrialization', in S. Conti, E. Malecki and P. Oinas (eds), *The Industrial Organisation and Its Environment: Spatial Perspectives*, Avebury, Aldershot, pp. 99-122.

Todtling, F. (1994), 'The uneven landscape of innovation poles: Local embeddedness and global networks', in A. Amin and N. Thrift (eds), *Globalization, Institutions, and Regional Development in Europe*, Oxford University Press, Oxford, pp. 68-90.

Zarhi, S. and Achiezra, A. (1966), *The Economic Conditions of the Arab Minority in Israel*, Arab and Afro-Asian Monograph Series, No. 1, Centre for Arab and Afro-Asian Studies, Givat Hviva.

Chapter 14

Enterprise, Embeddedness and Exclusion: Business Relationships in a Small Island Developing Economy

Michael Taylor

Introduction

The purpose of this chapter is to explore the nature of inter-firm relationships in the small developing country economy of Fiji to reflect on the complex processes of embedding that are currently held to underpin successful local economic growth in developed and developing countries alike. The model of embedded local growth, built on the burgeoning literature on industrial districts (Humphrey 1995), has at its heart mechanisms of *inclusion* based on trust, reciprocity and loyalty amongst small and medium sized enterprises (SMEs) in a place. It is argued that international competitiveness in a globalising world can be achieved through local collaboration and cooperation between firms that builds local knowledge economies that are innovative and dynamic (Maskell *et al.*1998, Maskell and Malmberg 1999, Lundvall and Johnson 1994, Asheim and Isaksen 2000). Through these localised mechanisms of cognitive, cultural, political and structural embeddedness (Polanyi 1957, Granovetter 1985, Grabher 1993), social capital is said to be generated (Putnam 1993) which acts as a public good capable of building and sustaining internationally competitive local economic growth (Storper 1997, Cooke and Morgan 1998).

The argument of this chapter is that this model has been inappropriately universalised to embrace Third World economic growth and the local, host country engagement of TNC subsidiaries. It is contended here that this simple, *inclusionary* model ignores significant *exclusionary* tendencies involved in embedded inter-firm relationships, especially when inter-firm relationships are orchestrated by transnational corporations. Thus, local embeddedness might be divisive and exclusionary, with the social capital that is said to be created through collaboration, trust and reciprocity, being less of the pure public good it has been argued to be. These exclusionary tendencies are implicit in analyses of power inequalities in networks of embedded firms in developed countries (Dicken and Thrift 1992, Grabher 1993, Taylor 1995, 2001), but they are thrown into stark relief in the developing country context – especially in Fiji.

The discussion of the chapter is developed in two stages. First, the developed country variant of the embeddedness model of local growth is elaborated together with the Third World and TNC extensions of the model. Here the *inclusionary* arguments of the model, built on issues of trust, reciprocity and loyalty, are tempered with counter tendencies towards *exclusion* that are evident in the same relationships. Second, these inclusionary and exclusionary tendencies are elaborated in the concrete circumstances of inter-firm relationships within the formal economy of Fiji. In this analysis three main types of enterprise are recognised: foreign-owned firms, Indian-owned family business networks, and small, single plant, livelihood firms (principally Indian-owned but including the small number of businesses owned by native Fijians) that are principally concerned with 'coping'.

'Inclusion', 'Exclusion' and Embeddedness

Clustering, Learning and Innovation: The First World Model of Local Growth

In the past decade, a powerful model of local economic growth has developed that draws on a range of complementary literatures on new industrial spaces, learning regions, innovative milieu, regional innovation systems and clustering (Braczyk *et al.* 1998, Maskell *et al.* 1998, Porter 1998, Simmie 1997, Storper 1997, Taylor and Conti 1997). At the heart of the model are complex processes of 'embedding' that recognise the enmeshing of firms' economic relationships within the broader social structures and social relationships of a place. Embeddedness has been recognised as assuming a range of forms (Grabher 1993). *Cognitive embeddedness* pinpoints firms' bounded rationality and place-based knowledge. *Cultural embeddedness* highlights the collective understandings they have on the way business is done in a place. And *political embeddedness* focuses on the place-specific struggles that firms have with non-market institutions. Structural embeddedness refers to the social dimensions of firms' transaction structures and their incorporation into networks of economic, social and cultural relations built on trust, reciprocity and loyalty. In clusters where these structural relationships prevail, tacit knowledge is mobilised, blended with codified knowledge (Asheim and Isaksen 2000), and spread through mechanisms of sharing rather than appropriation (Leborgne and Lipietz 1992). The result is local invention and innovation, building islands of sustained local accumulation with globally superior levels of productivity (Porter 1998).

At their core, the socially based processes of embedded local economic growth in this model are *inclusionary*, binding a suite of enterprises (usually SMEs) into a collectivity that generates social capital. That social capital is created in at least four ways: by 'value introjection' (shared moral and social values); by 'reciprocity transactions' (favours and approvals at an individual level); by 'bounded solidarity' (for example, class consciousness); and by 'enforceable trust' (compliance in the anticipation of reward) (Portes and Sensenbrenner 1993). The same processes also create 'institutional thickness' that further bolsters local economic growth processes (Amin and Thrift 1994, 1997). The global mosaic of these successful industrial spaces is, in turn, orchestrated by TNCs, global capital and global political institutions.

This institutionalist model tends to suggest that cooperation and collaboration between firms is the only mechanism that can create economic growth in localities confronted by globalisation. Cooperation is, in effect, elevated beyond market mechanisms as the principal source of competitive advantage. Trust is more important than price in shaping inter-firm transactions.

There is, however, substantial research that suggests that embeddedness and social capital are no substitute for market mechanisms. Uzzi (1996, 1997) has shown in the context of the New York garment industry that trust and reciprocity in transactions need to be balanced by considerations of price. Trust, loyalty and repeat business let firms tap into the local knowledge networks. At the same time, price-based transactions feed market signals into those same firms, keeping them sensitive to issues of costs, revenues and profitability.

But, economic relationships involve not just collaboration and competition, they also involve the exercise of power, and that is altogether more brutal and *exclusionary*. Power in buyer-supplier transactions is exercised through the intentional control of resources, the manipulation of relationships and the imposition of discipline (Taylor 1995, Wrong 1995, Allen 1997). Power inequalities restrict firms' business opportunities and their capacity to accumulate capital. They create lock-ins and create uneven spatial effects and they reserve strategic decision-making to specific social elites (especially in large corporations) (Taylor 2000, Hill 1995). However, these exclusionary aspects of the embeddedness model of local growth remain to be fully explored in analyses of industrial clusters in developed industrial economies.

Two Extensions of the Embeddedness Model

Two variants of this model have also emerged extending the notions of embeddedness and the creation of social capital through 'inclusionary' processes to embrace economic development in the Third World and to reinterpret the actions of TNCs. But, again, issues of 'exclusion' are pushed into the background (Putnam 1993, Yeung 1998a, 1998b).

In the developing country context the model is transformed from a *commentary on potential local growth processes* in the developed world, into a *prescription for actual local growth* in the developing world. Here it is suggested local communities need to be mobilised. Fostering local cooperation between firms and within communities and establishing local institutions can, by this reasoning, create globally compatible local economic growth with the very lightest touch of government. This institutionally driven research is relatively new (Schmitz and Musyck 1993). The major lending institutions, however, have recently been pushing for good governance to create an enabling environment for economic growth in the developing world, so it can be assumed that such studies will be of growing importance as these policy initiatives mature (Fine 1999, Mohan this volume).

Empirical research is less supportive of this view of embedded economic growth in developing countries (see Humphrey 1995). In both Brazil and India, for example, external competition, the activities of large corporations, and pre-occupation with exports have been shown to break down reciprocal relationships based on trust. They have forced firms to interact largely on a cost/price basis, destroying the autonomy of small firm subcontractors and intensifying labour exploitation (Schmitz 1993). The conclusion to be drawn from these empirical analyses is that *inclusionary* processes of reciprocal exchange are only weakly developed in industrial clusters in many developing countries (see Humphrey (1995) and the special edition of *World Development*). Instead, it would appear that *exclusionary* processes, orchestrated by large corporations, lead to the appropriation of clusters' externalities and transform cooperative growth into exploitation. What we do not know, however, is how extensive those *exclusionary* processes are in the global periphery.

TNCs and Embeddedness: Sleeping with the Enemy

The elaboration of the embeddedness model to include the enmeshing of transnational corporations (TNCs) within local economic networks of trust, reciprocity and loyalty is arguably the most contentious and problematic extension of this set of ideas. That TNCs grow out of particular local contexts and acquire the characteristics of those places through the complex processes of networking and embedding that create them is quite unproblematic (Dicken and Thrift 1992). However, the argument that the subsidiaries and overseas operations of those corporations can become embedded within the local networks of the host countries within which they operate is far more contentious. Now it is suggested that it is both possible and desirable for TNCs' subsidiaries simultaneously to be reciprocating, local and trusted members of local host country networks while complying with and contributing to the corporate strategies of their parents (Yeung 1998a).

Certainly, as TNCs expand abroad they carry the 'flavour' of their local origins with them (Pauly and Reich 1997, Yeung 1998a). Now it is being suggested that the subsidiaries of TNCs must be locally embedded in their host countries in order be successful. To gain place-specific knowledge in their host countries it is argued that TNCs need local experience and, therefore, have to be less 'foreign'. From an extensive literature review, Yeung (1998a) has suggested that, in the world's principal financial centres and in manufacturing centres in the US and Mexico, for example, TNC subsidiaries can only benefit from place-based externalities when they are embedded in complex local, social and personal relationships.

By inference, therefore, subsidiaries of TNCs can magically become good local team players at the heart of local inclusionary processes. This is an interpretation of the activities and actions of TNCs that is fundamentally at odds with the findings of decades of research. That research has almost universally interpreted the overseas operations of TNCs as exploitative and exclusionary. Exclusion is at the core of the New International Division of Labour thesis, theories of multinational and transnational corporate development (Taylor and Thrift 1986) and the decades of critique of US TNCs (Taylor 1999). This longstanding critique cannot be ignored but deserves to be reasserted.

The Fiji Economy: Segmentation, Race and Development

The remaining sections of this chapter explore the balance of these countervailing processes of inclusion and exclusion associated with the different types of enterprise that operate within the small island developing economy of Fiji. Fiji is a Pacific Islands republic of some 780,000 people. It has a volatile racial mix in which Fijians make up 51% and Indo-Fijians (descended mainly from indentured labour) 44%. The backbone of the economy is agriculture, fisheries and forestry, especially sugar production, but the value of agriculture as an export earner and contributor to GDP is declining. Increasingly important are manufacturing, trade and hotels, and transport and communication. But these formal sector activities create only about 1500 jobs a year. Agriculture already employs 44% of the economically active population in formal sector operations and subsistence, and absorbs the remainder of the 10,000 annual increase in the labour force (Chandra 1998). The formal sector is capable of creating many more jobs. Tax Free Factories (TFF) producing garments, for example, have created some 15,000 jobs since 1988. But, the military coups of 1987 have led to the massive emigration of Indo-Fijians and their capital. Indo-Fijians dominate business ownership in the formal sector but they are now justifiably wary of investing. Native Fijians, in comparison, are more rarely business owners (Qalo 1997). They tend, instead, to remain on the land or to take up public sector jobs. But those public sector jobs are being shed as government deregulates, corporatises and privatises to contain public spending. This cocktail of forces is an explosive mix, exacerbating Fiji's racial problems.

Against this background, capitalist development in Fiji has created a range of local circumstances that have spawned very different types of formal sector business enterprise. For the present study, three separate eras can be distinguished each of which has spawned a distinctive assemblage of enterprises:

- The colonial era (1874 to 1970) of plantation production that was served by large foreign-owned trading and transport companies and small, domestic livelihood enterprises in retailing, services and small scale production;
- The import replacement era (1970 to 1987) which heralded the entry of more foreign firms, and the emergence of large-scale Indo-Fijian business, frequently organised as family networks; and
- The era of export orientated deregulation (1987 on) which principally has seen the rise of Tax Free Factories in the garment industry and the creation of large numbers of jobs outside agriculture.

The history of economic development from cession to Britain in 1874 to independence in 1970 was one of colonialism and a colonial division of labour. Fiji provided Britain with primary and agricultural products (principally sugar and copra) from large scale plantations operated by indentured labour brought from India. Significant tracts of the best agricultural land were leased in small lots to Indian cane farmers. Ownership, however, remained with the indigenous Fijian, contributing to the racial tensions that have erupted in 1968, 1987 and

2000. Local processing of this produce was minimal. Sugar and copra milling, '... were not designed to make Fiji a manufacturing country; they were designed primarily to make Fiji a better agricultural country, by making exports easier – a common objective of metropolitan countries in their colonies' (Chandra 1998, p. 114). Trade and agricultural processing were in the hands of large colonial companies – including Burns Philp, Morris Hedstrom, WR Carpenter and Colonial Sugar Refiners (CSR), for example – that were always close to the colonial government. Now, much changed and under new ownership, these companies are the historic core of foreign-owned business in Fiji.

After independence and with the adoption of import-substitution policies to build and diversify the economy into manufacturing, many more TNCs and foreign firms (especially from New Zealand and Australia) began to operate in Fiji. They came to dominate not only trade and production but also resorts and hotels as the tourist industry grew. Import substitution also encouraged the establishment of many small domestic enterprises. Most frequently, these were low-skilled livelihood enterprises. Almost exclusively they were Indian-owned, as the descendants of indentured labourers, precluded from owning agricultural land, sought advancement through commerce. Some of those Indian-owned domestic companies expanded rapidly to become extensive, often volatile, family networks. Those family networks include, for example, the Punja group, G.B. Hari, Manubhai, Motibhai, Rup, and the Reddy group of companies. Import substitution policies necessarily draw business and government close together, and politics have been and continue to be vital to the operations of both these family networks and foreign firms.

Following the racially motivated coups of 1987 when, as again in 2000, native Fijians chose violence to wrest political control from an elected government led by an ethnic Indian, the economy was in danger of collapse. Indo-Fijians, Indo-Fijian business and Indo-Fijian capital fled the country. Some stayed and prospered in an environment of reduced competition. The Rambuka government embraced economic deregulation. This was not for any ideological reason but purely to create urgently needed jobs and to cut government spending. Trade deregulation was initiated, a Tax Free Factory scheme was introduced in 1988 to stimulate manufacturing exports, the tax system was reviewed and restructured, and a start was made on corporatising and privatising government departments and statutory organisations (Chandra 1998). Deregulation and new concessions transformed the business environment and increased competition. Some foreign firms pulled out, local manufacturing was forced to adjust, and new business opportunities were created. It was new foreign firms (for example Ghim Li from Malaysia), foreign investors (for example from Korea) and Indian family businesses (for example, the Solanki family) that capitalised on those opportunities.

Through these phases of development three main types of enterprise have emerged: foreign-owned firms, Indian-owned family business networks, and small, single plant, livelihood firms. Each performs a very different role in the Fiji economy. The remaining sections of this chapter will explore the nature and form of the embeddedness of these types of enterprise within the economy to reflect on their potential to create local growth.

The Embeddedness of Foreign Firms in Fiji

Foreign-owned firms have very distinctive relationships with other Fiji-based enterprises and the Fiji economy in general. Many of these ventures were established in the era of import-substitution industrialisation to mop up the local Fiji market and to develop exports into the neighbouring Pacific Islands (Taylor 1986). They were always close to government, especially before independence in 1970, and they were not averse to showing their 'gratitude' to government after independence. Some are part of a local commercial mythology, especially 'Our' Burns Philp (as it once styled itself) and Carpenters. They were pillars of the commercial community: Australian-owned until over-extension in the US of their parent companies led to their sale to Malaysian and British interests. Confronted by deregulation in the past 13 years foreign firms have left, have been sold and have changed their relationship with the economy. Now they are concerned with secure rates of return. Their embeddedness in the Fiji economy is increasingly selective. Arguably, it is now more exclusive than it is inclusive.

First impressions are that the foreign firms surveyed in December 1998 demonstrate significant local embeddedness. Their strategy is one of 'getting close to customers', of being, 'their strategic partners; [to] get business from major customers and to meet the competition locally and from overseas' (firm S9). 'The main thrust of what we're doing at present is trying to build relationships with [customers] and trying to be the single source to solve all their problems' (firm S15). A TNC engaged in retailing (firm S14) has 'very loyal relationships' with its 15 or so Fiji-based suppliers – 'it's kind of interdependence you know. Like we depend on them for our supplies and, because most of them are not really large ... the size of their operations justifies keeping to us. The thing about having a strong bond is that they are dealing with us all day every day.'

Some inter-firm relationships have all the hallmarks of embeddedness, involving trust, loyalty and reciprocity. As one respondent explained:

Oh well, there's finance. We can offer safety advice and training. We can do workshop audits. We can do health and safety training because we have fairly advanced systems in a lot of these areas. And then we can get help from any country in the world to help our customer with a specific problem. (firm S15)

In the same vein, another respondent (firm S9) explained that it sent samples of a production residue from one of its major client's processing plants to New Zealand for analysis, at no cost to the client. Equally, a food manufacturer relied on suppliers' technical backup, 'because we don't have very good technical expertise locally' (firm S13). A cleaning products producer and a retailer (firms S7 and S14) nurtured selected local suppliers principally by supplying technical advice and guaranteeing sales. Indeed, they helped to create those local businesses. Interviews with furniture subcontractors elaborated these relationships. A furniture manufacturer (firm S11) explained that they had been helped by being supplied, 'with orders for goods and items, and paying on time' and, when the business was being started, by being supplied with 'imported materials like leather and fabrics' to work on. They also received technical advice. The relationship was, in turn, reciprocated by the suppliers. They anticipated periods of high demand at Divali and Christmas by making for stock without orders (firm L5). However, both small subcontract suppliers pointed out that they had begun as general furniture manufacturers in the 1970s but had specialised at the retailer's request. They had certainly prospered, but because of the retailer's policy of cycling expatriate buyers through Fiji and the other countries it works in, they experienced periodic disruptions as new buyers acclimatised to local business conditions.

It is clear, nevertheless, that only a very few local firms have prospered as subcontract furniture suppliers to this particular retailer: by one account only five – Rup Investment, Rup Industries, Popular Furniture, Popular Industries and Comfort Home Furnishing. A 1983 survey had revealed many more furniture subcontactors. But, as the retailer (firm S14) explained, suppliers had been dropped after the 1987 coups, 'because they weren't big enough and there wasn't enough market' and because, euphemistically, 'some sort of packed up and went away' (fled the threat of racial violence).

What is clear, however, is that most foreign firms build close buyer-supplier relationships with only a select group of counterparts, who are usually (though not always) the biggest and the most influential firms. '80% of our business is with our 20 largest customers, and they are who we want to get close to' (firm S9). 'We go through phases in business development and our current phase, if you like, is moving back to looking after these key customers with key sales people. We've tried in the past to be all things to all people and,

I think ... this swing of the pendulum we've decided we can't do that' (firm S15). For foreign firms, closeness is also used to ensure payment, 'so they make prompt payments ... so they just take 30 days and they pay it on 30 days' (firm S3). Without exception, all ten foreign-owned firms interviewed will sell to local, small and livelihood firms at best on 30 days terms but almost always only for cash. Trust and reciprocity does not figure in these relationships. The foreign-owned firms are not in the business of creating local social capital and acting as catalysts in local knowledge networks. Instead they exclude or quarantine local, small firms whose transactions are risky, in an effort to guarantee their own revenues and to meet the financial targets set overseas by their parent organisations.

It would seem, therefore, that the current style of TNC business is to get 'up close and personal' with buyers and suppliers in host countries, but with only the few, bigger firms. Embeddedness in this sense is to guarantee markets, market share and production, with production risk minimised through minimal financial commitment, and local market risk minimised by dealing with only the least risky clients. It is nothing to do with the creation of knowledge, learning and innovative growth. It is the hard edge of rational capitalism with a major element of ephemerality and opportunism, certainly in the garment sector. TNC embeddedness in Fiji is demonstrably more a mechanism of exclusion than a mechanism of inclusion. It is a mechanism of selective collaboration to secure production and market share that creates minimal social capital.

Embeddedness and Indo-Fijian Family Business Networks

Developing alongside foreign-owned businesses in Fiji, especially during the era of import substitution policies, is a set of Indo-Fijian businesses whose members have expanded to become major national concerns, and the principal source of domestic enterprise. Some are multi-site, multi-sector operations and, in the 1980s, at least one made tentative offshore moves (Taylor 1986). More recently, their ranks have been swelled by similar enterprises that have grown in the era of deregulation and Tax Free Factories. What is distinctive about the engagement of these enterprises with Fiji's economy and society is:

- their embeddedness within extended family structures;
- their embeddedness within their ethnic communities, especially through philanthropy; and
- their embeddedness within the political structures of the country.

Strong family networks are at the heart of large Indo-Fijian businesses. Some are very large such as the Punja group with 12 subsidiaries in a wide range of manufacturing and retailing activities. Others are much smaller like the Chand family with brothers making concrete products in Suva and Lautoka and a nephew setting up distribution depots in Nadi, Sigatoka and Lautoka. The networks cross a wide range of sectors. The Reddy group is diversified with large interests in hotels and resorts, construction, property, insurance and other enterprises. Rup family interests span furniture making, construction and retailing. Motibhai retails hardware, as does the Vinod Patel group which also makes wire products. The Manubhai group retails building materials and hardware. The Hari family was once heavily engaged in retailing, but now concentrates on property investment and garment manufacturing. Members of the Solanki family are also engaged in garment manufacturing for export.

Family ties have been vital to the building of these networks, and strong family ties are a feature of the Indo-Fijian community. Members of the extended families not only own, manage and run the businesses, they staff them too. For example, the Prasad family from Lautoka, who own the Foodtown grocery franchise and operate 7 stores across the country, had 425 employees in 1995, 150 of whom were family members. In virtually all these groups, members of the younger generation are groomed to assume the mantle of control, and the generational links can also be extremely strong. Vinod Patel, the head of the hardware retail group that carries his name, was quoted in 1996 as saying:

> The family should be united, work as a team. We want to make solidarity for the next generation so that it works the same way. My thinking is more for family than money. (Arun 1996, p.67)

This is a widely reported sentiment among Indo-Fijian business. At the same time, however, these family groups are very fragile and break up easily and acrimoniously through family disagreements, especially when they are coupled with generational changes. This happened to the Popular group of companies which is now split into Popular Industries and Popular Furniture. It happened to the Rup family in 1993: now Rup Industries and Rup Investments (including Rup Big Bear stores). And it happened to the Solanki family when the two brothers running the business parted company in 1984.

The social ties of these businesses go beyond family to embrace the ethnic Indian community from which they have sprung. It is common for successful businessmen heading family groups to be religious, and philanthropic within their ethnic community. They tend to give to religious, welfare and educational foundations, strengthening community bonds in ways that have been identified as creating social capital. With benefactors and role models it is easy to see why the Indo-Fijian community continues to generate entrepreneurs.

Equally vital to the success of these networked family enterprises have been strong political ties, both direct and indirect. Notwithstanding the built-in bias of the political system in favour of native Fijians, which was cemented in the now defunct, racist Constitution of 1990 (which leaders of the armed 2000 uprising want reinstated), it is important for Indian business to be close to government. Thus, Hari Punja, patriarch of the Punja group, is a member of parliament. And Y.P. Reddy, head of the Reddy group,'used to be a key Indian member of Mara's [now President Ratu Sir Kamisese Mara] old party, the Alliance, which ruled Fiji for 27 years and then collapsed because of an army coup. Now he is president of the National Federation Party, the main Indian party and the Alliance's old enemy' (Keith-Reid 1995, p.30). Ramesh Solanki, head of one arm of the Solanki family, was a private sector representative at the SPARTECA talks in 1982 that negotiated duty free access for Pacific Islands products into Australia and New Zealand (Naidu 1994). He was the first to export garments from Fiji under the agreement, and separately his brother Ranjit was also one of the first to exploit the opportunity. Then, in the late 1980s, after the 1987 coups, 'He was instrumental in supporting the introduction of the Tax Free Zone/Tax Free Factory scheme' (Singh 1998 p. 30). He set up Classic Apparel in 1987 with Minister Jim Ah Koy and the Stafford Group of Australia, and the scheme became operational in 1988. In 1989, he set up United Apparel, which now employs over 1000 people, and he went on to help establish another garment factory, Farah (Fiji) Ltd. Sometimes, however, business links with government become a little too close. In 1997, a former Fisheries Minister Ratu Ovini Bokini was charged with 32 counts of official corruption relating to alleged bribes received from the country's then premier fishing company, Fiji Fish (Foster 1997).

Tax free factory status has now created a new form of non-local embeddedness among some of the larger domestic enterprises – strong, highly selective, and ephemeral structural embeddedness with corporate and TNC clients *outside* Fiji. Most garment manufacturers have one principal client (firm L10) and work principally on a 'cut, make and trim' basis, rather than more sophisticated forms of working (firm L1). They work closely with those overseas clients to develop products (firm L1). They are supplied with materials (firm S16), consultants (firm S19), and sometimes with technology. As it was explained concerning links with one Australian client:

> So they came here, ... so all the technology, whatever they had, they've given to us because they had a factory already. [They supplied] machinery, advice,

training, everything. They were closing down and the reason they gave it us was to make sure they get the right quality and everything, same standard. (firm L10)

But, if Fiji garment manufacturers increase efficiency too far, through computer pattern-making or computer cutting, they might cease to qualify for the 50% local content required for access to the Australian market under the SPARTECA agreement (Keith-Reid 1995, p. 33). This shows how fragile this overseas structural embeddedness is, as large Indo-Fijian garment firms begin to compete directly with producers in China, the Philippines and Sri Lanka.

What is clear then is that Indo-Fijian business has used family embeddedness as a coping strategy to compete with the large foreign interests that dominated the Fiji economy for decades. That social embeddedness has extended to the wider Indian community. However, though effectively second-class citizens in Fiji, the heads of Indo-Fijian business networks have had to become politically embedded with the politically dominant native Fijians to gain the institutional support they need to be commercially successful. Now, export orientated industrialisation is adding a new dimension to these relationships, as large domestic garment manufacturers become structurally, but ephemerally, embedded with overseas corporations through their buyer-supplier transactions.

The Embeddedness of Domestic SMEs

In the domestic small firm sector, among Indo-Fijian SMEs, price issues rule buyer-supplier relationships, with the smallest enterprises operating in an essentially cash economy. The trust and reciprocity that is reckoned to create social capital is almost completely absent. For many firms this is the local way of doing business in Fiji. Relationships with other firms are often long-standing but are only ever price-based. Embedded relationships amongst small firms are very much the exception rather than the rule. For example, a small engineering firm (firm L2) offers only 15 of its 100 clients 30-day terms (though they frequently take 60 days to pay). The rest pay cash. Another (firm L4) has only two account customers, '... and all the small ones we give three or four days or five days, ten days, like that. And sometimes we get it cash, about $40, $50, $60 like that'. The firm itself has to pay cash for all its own purchases while one of its clients, the Fiji Sugar Corporation, always takes 60 days trade credit! A small snack foods manufacturer (firm S5) adjusts packet weights to keep prices stable. 'Well, it's just a market basis' (firm S8), and one in which the small are squeezed. 'Here, ... because people know us and they know us too personally, they want it [goods] dirt cheap, you know - the sale price' (firm L6 making garments for the local market).

Firms are also played off against each other in this section of Fiji's economy.

> They'll use you here. They buy on account from us, run up $5000 and just shoot off and go to the other guy, our competitor, and he (sic) buys from them. And he's not paid us in say 40 days and it's a 30-day account, and he expects us to give it to him. (firm S17, making building materials)

Indeed, throughout the interview with one small electrical contractor (firm L9) the manager was nothing short of obsessive about paying and being paid within 30 days.

Price is certainly more important than quality in this section of Fiji business, and this was remarked on by a number of firms. To quote the owner of a small engineering firm (firm S18):

> Fiji at the moment is very price orientated. So much so that customers will ask you to do substandard quality work to get it within the price. And daily you will have a fellow come in and say 'Look, I just want this to go another two weeks, two months or whatever, so just do it up to keep it going for that long'. Sometimes the clients ask you to compromise quality.

Deeper relationships are being built by some domestic SMEs and amongst the survey firms these tended to be the larger, longer established businesses. As it was explained by the director of a domestic engineering firm:

> We have very few occasional customers. Whoever we deal with we have long-term relationships with. I mean you do also develop personal relationships with your customers -- you've got to mix a bit of pleasure with it too. Financially, we have to carry customers at times. I mean, strictly speaking in accounting terms, it's meant to be 30 days trading terms. But, I mean, most of them don't actually respect that and they have, you know, problems, and then get approached for us to extend the terms. And we do, in consideration of the business that we have conducted with them over the years. (firm S6)

From a small maintenance engineering firm's perspective, '[a large local manufacturer] is one company that give (sic) me job and I quote. And if they think it's too high or something they will come back to me and ask me to look again, look at my pricing and everything. But they stick with me' (firm S4). This same firm, along with several others, pointed out that to win tenders or to have quotations accepted may mean giving 'back-handers' to the purchasing staff of client firms, and that this was not an unusual practice in Fiji.

The embeddedness that there is among this set of businesses is more to do with coping than with knowledge networks and innovation. A small garment manufacturer (firm L7), established for less than a year and hoping to export, illustrates the situation. The business has only price-based cash relationships with its clients, and produces on a 'cut, make and trim' basis. The Indo-Fijian owner started the business after being an investor in another tax-free factory. He went into business with a nephew, and now family members make up 30% of the workforce. A friend provides occasional subcontract work, and the business in which he was formerly a partner channels T-shirt and baseball cap work for local resorts to him. Adding to the web of support, equally new neighbouring garment firms lend him machines for a day or two when his own break down.

The picture that emerges, therefore, of buyer-supplier relationships amongst the smaller, Indo-Fijian firms is that trust, reciprocity and loyalty are very weakly developed. This is no embedded network of information and knowledge transfer that creates globally competitive local growth. This is price-based, short-term business in which quality is compromised and graft is not uncommon. The limited embeddedness that does exist in no way constitutes a competitive strategy to gain market share in a global economy. Embeddedness here is a *coping strategy* to make a living in a hostile and unstable domestic environment in which foreign firms and large local firms seek either to exclude them or exploit them.

Conclusion

The purpose of this chapter has been to explore the nature of enterprise embeddedness in Fiji, to more fully come to grips with the complex social underpinnings of economic relationships within that country and its communities. The currently popular interpretation of those relationships are guided by the embeddedness model of local economic growth. This is an institutionalist interpretation of socially based economic relationships that highlights the importance of trust, reciprocity and loyalty, rather than competition, market processes and the exercise of power, as the basis of growth. It is an 'inclusionary' model that has been extended to include developed and developing country contexts, and to incorporate TNCs as collaborative local contributors rather than externally controlled exploiters.

Even at a theoretical level, it was suggested that this model and its variants neglect significant 'exclusionary' processes that restrict growth and run counter to the 'inclusionary' processes they promote. However, the empirical evidence from Fiji demonstrates that even in a small developing economy the mix of 'inclusionary' and 'exclusionary' social processes is far more complex than theory would suggest.

In this racially divided society, three types of enterprise are identified, each of which is very differently embedded in the local economy. Foreign-owned firms operate in Fiji to mop up the local market not to develop an export platform. They are, in effect, selectively embedded. They are locally embedded only in circumstances where risk is minimal and revenues are assured. This means they are structurally embedded with the 20% of clients that generate 80% of their revenues – the large domestic and other foreign firms. Exclusion is a key component in this form of embeddedness. Growth spin-offs are limited and controlled. They are not the externalities and pure public goods of theory. The 'excluded' enterprises are the domestic Indo-Fijian SMEs and micro enterprises that the embeddedness model proposes as a source of social capital and the engine of local growth. They are condemned to a short-termist, cash-only economy in which trust, reciprocity and loyalty is in short supply. Here is a commercial arena of compromised quality and graft, driven by price. Social, cultural and structural embeddedness is minimal, but where it exists it is more a coping mechanism than an engine of growth.

In contrast, the most strongly embedded enterprises in Fiji are the Indo-Fijian family networks. But here, the strongest aspects of embeddedness are not structural – they are not the reciprocal, trusting relationships that create local 'learning' economies and the forms of social capital these bring. Instead, embeddedness is strongly social and cultural and focused on the family and the ethnic Indian community. Possibly of more significance, these enterprises are strongly politically embedded. Commercial opportunities come through political channels as much if not more than through the structural buyer-supplier ties that lie at the heart of the embeddedness model of local economic growth. Politically guided commercial opportunities are by their very nature exclusive, and not far below the surface in Fiji is the issue of cronyism. By no stretch of the imagination can these circumstances in Fiji create social capital. They can only thwart it. Indeed, it is in this political arena that the elite of the native Fijian community exercises its economic power. Few native Fijians are engaged in private sector enterprise and Qalo (1997) has only been able to identify 100 Fijian-owned businesses. Instead, the Fijian elite controls the policy levers that determine the issue of licences, set tariffs and quotas, provide loans and so on. This is their point of entry into local processes of capital accumulation.

What this empirical analysis has shown is that embeddedness processes, even in a small country, are complex and interactive. They are inclusionary and exclusionary at one and the same time. They divide and exclude as much as they combine and unite. They are not the naively conceived mechanism of theory that creates social capital and growth. Clearly, there is an urgent need to more fully unpack embeddedness as a concept, to appreciate its social and cultural specificities and to come to grips with its less benign aspects.

Acknowledgments

The author gratefully acknowledges the financial assistance of the Nuffield Foundation, who funded the travel to Fiji, and the business owners and managers in Fiji, who gave freely of their time.

References

Allen, J. (1997), 'Economies of power and space', in R. Lee and J. Wills (eds), *Geographies of Economies*, Arnold, London, pp. 59-70.

Amin, A. and Thrift, N. (1994), 'Living in the global', in A. Amin and N. Thrift (eds), *Globalization, Institutions and Regional Development in Europe*, Oxford University Press, Oxford, pp. 1-22.

Amin, A. and Thrift, N. (1997), 'Globalization, socio-economics, territoriality', in R. Lee and J. Wills (eds), *Geographies of Economies*, Arnold, London, pp. 147-157.

Arun, N. (1996), 'Between the good and the best', *The Review, Fiji*, July, pp. 64-67.

Asheim, B. and Isaksen, A. (2000), 'Localised Knowledge, Interactive Learning and Innovation: Between Regional Networks and Global Corporations', in E. Vatne and M. Taylor (eds), *The Networked Firm in a Global World: Small Firms in New Environments*, Ashgate, Aldershot, pp. 163-198.

Braczyk, H-J., Cooke, P. and Heidenreich, M. (eds) (1998), *Regional Innovation Systems*, UCL Press, London.

Chandra, R. (1998), 'Industrialisation', in R. Chandra and K. Mason (eds), *An Atlas of Fiji*, Department of Geography, School of Social and Economic Development, The University of the South Pacific, Suva, pp. 114-117.

Cooke, P. and Morgan, K. (1998), *The Associational Economy*, Oxford University Press, London.

Dicken, P. and Thrift, N. (1992), 'The organization of production and the production of organization; why business enterprises matter in the study of geographical industrialization', *Transactions of the Institute of British Geographers, New Series*, vol. 17, pp. 279-291.

Fine, B. (1999), 'The developmental state is dead – long live social capital?', *Development and Change*, vol. 30, pp. 1-19.

Foster, S. (1997), 'Fishing with the enemy', *The Review, Fiji*, February, pp. 12-19.

Grabher, G. (ed) (1993), *The Embedded Firm: On the Socioeconomics of Industrial Networks*, Routledge, London.

Granovetter, M. (1985), 'Economic action and social structure: a theory of embeddedness', *American Journal of Sociology*, vol. 91, pp. 481-510.

Hill, S. (1995), 'The social organization of boards of directors', *British Journal of Sociology*, vol. 46, no. 2, pp. 245-278.

Humphrey, J. (1995), 'Introduction', *World Development*, vol. 23, no. 1, pp. 1-7.

Keith-Reid, R. (1995), 'A Strong Voice on the Investment Problems', *Islands Business Pacific*, Special Report, July, pp. 25-40.

Leborgne, D. and Lipietz, A. (1992), 'Conceptual Fallacies and Open Questions on Post-Fordism', in M. Storper and A. Scott (eds), *Pathways to Industrialization and Regional Development*, Routledge, London, pp. 332-348.

Lundvall, B.Å. and Johnson, B. (1994), 'The learning economy', *Journal of International Studies*, vol. 1, no. 2, pp. 23-42.

Maskell, P., Eskelinen, H., Hannibalsson, I., Malmberg, A. and Vatne, E. (1998), *Competitiveness, Localised Learning and Regional Development. Specialisation and Prosperity in Small Open Economies*, Routledge, London.

Maskell, P. and Malmberg, A. (1999), 'Localised learning and industrial competitiveness', *Cambridge Journal of Economics*, vol. 23, pp. 67-190.

Naidu, R. (1994), 'The Shackles of SPARTECA', *The Review, Fiji*, June, pp. 32-46.

Pauly, L. and Reich, S. (1997), 'National structures and multinational behavior: enduring differences in the age of globalization', *International Business*, vol. 51, pp. 1-30.

Polanyi, K. (1957), 'The economy of instituted process', in K. Polanyi, C. Arensberg and H. Pearson (eds), *Trade and Markets in Early Empires*, Free Press. Glencoe IL, pp. 243-270.

Porter, M. (1998), *On Competition*, Harvard Business School Press, Boston MA.

Portes, A. and Sensenbrenner, J. (1993), 'Embeddedness and immigration: Notes on the social determinants of economic action', *American Journal of Sociology*, vol. 98, no. 6, pp. 1320-1350.

Putnam, R. (1993), *Making Democracy Work: Civic Traditions in Modern Italy*, Princeton University Press, Princeton NJ.

Qalo, R.R. (1997), *Small Business: A Study of a Fijian Family. The Mucunabitu Ironworks Contractor Cooperative Society Limited*, Mucunabitu Education Trust, Suva, Fiji.

Schmitz, H. (1993), *Small Shoemakers and Fordist Giants: Tale of a Supercluster*, IDS Discussion Paper No. 331, IDS, Sussex.

Schmitz, H. and Musyck, B. (1993), *Industrial Districts in Europe: Policy Lessons for Developing Countries?*, IDS Discussion Paper No.324, IDS, Sussex.

Simmie, J. (ed) (1997), *Innovation, Networks and Learning Regions?* Regional Policy and Development Series, Jessica Kingsley Publishers, London and Bristol PA and Regional Studies Association, London.

Singh, S. (1998), 'United Apparel: a win-win situation', *The Review, Fiji*, May, pp. 25-30.

Storper, M. (1997), *The Regional World*, Guilford Press, New York.

Taylor, M. (1986), 'Multinationals, Business Organisations and the Development of the Fiji Economy', in M. Taylor and N. Thrift (eds), *Multinationals and the Restructuring of the World Economy*, Croom Helm, London, pp. 49-85.

Taylor, M. (1995), 'The business enterprise, power and patterns of geographical industrialisation', in S. Conti, E. Malecki and P. Oinas (eds), *The Industrial Enterprise and Its Environment: Spatial Perspectives*, Avebury, Aldershot, pp. 99-122.

Taylor, M. (1999), 'The dynamics of US managerialism and American corporations', in D. Sadler and P. Taylor (eds), *The American Century*, Blackwell, London, pp. 51-66.

Taylor, M. (2000), 'The firm as a temporary coalition', Paper presented to the workshop on The Firm in Economic Geography, University of Portsmouth, March.

Taylor, M. (2001), 'Enterprise, Embeddedness and Local Growth: Inclusion, Exclusion and

Social Capital', in D. Felsenstein and M. Taylor (eds), *Promoting Local Growth: Process, Practice and Policy*, Ashgate, Aldershot, pp. 11-28.

Taylor, M. and Conti, S. (eds) (1997), *Interdependent and Uneven Development: Global-Local Perspectives*, Ashgate, Aldershot.

Taylor, M. and Thrift, N. (eds) (1986) *Multinationals and the Restructuring of the World Economy*, Croom Helm, London.

Uzzi, B. (1996), 'The sources and consequences of embeddedness for the economic performance of organizations: the network effect', *American Sociological Review*, vol. 61, pp. 674-698.

Uzzi, B. (1997), 'Social structure and competition in interfirm networks: the paradox of embeddedness', *Administrative Science Quarterly*, vol. 42, pp. 35-67.

Wrong, D.H. (1995), *Power: Its Forms, Bases and Uses*, Transaction Publishers, New Brunswick and London.

Yeung, H. (1998a), 'Capital, state and space: contesting the borderless world', *Transactions of the Institute of British Geographers, New Series*, vol. 23, pp. 291-309.

Yeung, H. (1998b), 'The social-spatial constitution of business organizations: A geographical perspective', *Organizations*, vol. 5, no. 1, pp. 101-128.

Chapter 15

The Local Embeddedness of Firms in Turkish Industrial Districts: The Changing Roles of Networks in Local Development

Ayda Eraydin

Introduction

In the recent literature on local development, a strong consensus has emerged that the global economy is constructed in and through localities with different characteristics and institutional capacities (Amin and Thrift 1995). The success of individual localities, it is claimed, depends mainly on process of local embeddedness involving the interaction of different local actors and institutions (Grabher 1993). Industrial districts, it is claimed, have a special position in this pattern of relationships because they possess advantages beyond the externalities and agglomeration economies of industrial clusters. Those advantages derive from the nature of inter-firm and inter-institutional relationships, collective understandings and actions, which create local social capital (Cooke and Morgan 1998, Maskell and Malmberg 1999a).

The advantages also include *untraded interdependencies*: the knowledge creation, reproduction and learning involved in socially constructed inter-firm relationships employing both tacit and codified knowledge (Malmberg 1996, Camagni 1991, Pyke and Senberger 1991, Belussi 1996, Storper 1995). They also include the social norms embedded in a district that constrain opportunistic behaviour and generate social capital (Amin 1999, Harrison 1994a, Angel 1991, Beccatini 1991, Morgan 1997, Asheim 1996, Maskell and Malmberg 1999b). The advantages of industrial districts are also said to derive from *institutional thickness* that is claimed to facilitate collaboration and generate innovation (Amin and Thrift 1994, Tödling 1994, Locke 1995, Rabelotti 1997, Gregersen and Johnson, 1997). All these characteristics are claimed to promote the interactive learning and innovation vital to sustain a district's international competitiveness (Malmberg and Maskell 1997, Amin and Cohendet 1999, Cooke *et al.* 1998).

Although the arguments developed in the literature suggest that local 'industrial atmosphere' might create growth, they also obscure the conflicts and tensions that exist in industrial districts (Brusco 1986, Staber 1996) and neglect the potential for local embeddedness to generate resistance to change and even delay change (Glasmeier 1994). Institutional thickness can act as a barrier as much as a stimulator of change (Raco 1999, Amin 1999), breeding domination, conflict and resistance though 'institutional overload' (Rabelotti 1997, Heidenreich and Krauss 1998, Glasmeier 1994). Indeed, local institutions may be incapable of solving a place's emerging problems (Schmitz 1998, Coriat and Bianchi, 1995). Equally, inter-firm relationships need not always be smooth and are just as likely to involve asymmetric power relations, exploitation and competition as trust and reciprocity (Taylor 1999). When localities face recession it is particularly difficult to sustain collaborative networks and to avoid power struggles (Glasmeier 1991, Cooke and Morgan 1994). There is also increasing evidence that locally embedded relationships that might have been very important in the initial phases of local growth might be less effective in later stages owing to increasing competition (Amin 1999). This situation raises questions about whether the advantages of embedded relationships in industrial districts are in fact temporary (Taylor 1999) and whether they are sufficient to cope with change.

This chapter examines the evolving importance of local embeddedness as it has changed through the phases of development of three industrial districts in Turkey – Denizli, Gaziantep and Çorum. It explores the dynamics and motivations of critical actors during the phases of emergence, take-off, upsurge and crisis in these industrial clusters, identifying the temporariness of some locally embedded relations and their inability to generate positive change during periods of economic downturn. The chapter also shows how both rapid growth and stagnation create power struggles and new coalitions, resulting in fragmentation and exclusion among local producers and the loss of economic coherence in these areas. It is concluded that aggressive survival policies shaped by global economic conditions can disadvantage weak groups, and that there is a need to more carefully evaluate the empirical detail of locally embedded inter-firm relationships.

Unexpected Growth: The Emergence of Industrial Districts in Turkey

In Turkey four metropolitan regions dominate industrial production and are the focus of increasing polarisation. They accounted for 67% of total manufacturing employment in 1971, and increased their share to 72.7% in 1995. Regional centres concerned mainly with domestic production have been virtually stagnant since the early 1970s and provide only 7.2% of total manufacturing jobs. Elsewhere, manufacturing has declined. In 1995, these other regions had 59.8% of the population (37.5 million people) but only 17.2% of manufacturing jobs. However, in sharp contrast, some localities within these less developed regions experienced rapid increases in manufacturing activity through the 1970s, 1980s and 1990s. Although it can be questioned whether these clusters constitute idealised 'industrial districts', their success in the last decade was very important in showing the economic growth potential of areas outside the major industrial concentrations in Turkey. Among these localities three have shown particularly strong growth in terms of numbers of jobs and numbers of production units; Denizli, Gaziantep and Çorum (Eraydin 1999). The trajectories of development in these three industrial clusters provide significant information on the role of embeddedness in promoting local growth processes.

These industrial districts are in different parts of Turkey. They are renowned for small business and network relationships between specialised, small, family enterprises. Around 85% of firms in these centres employ fewer than 100 workers and a high proportion of these small production units are family owned enterprises (in Gaziantep 86.1%, in Denizli 62.9%, and in Çorum 60.8%). Small firms dominated the growth of these districts in the 1990s. In Denizli, between 1990 and 1996, the numbers of establishments and employees increased 323% and 133.5% respectively. In Gaziantep, the rate of growth was lower: 185.8% for establishments and 63.8% for manufacturing jobs over a five-year period[1]. In Çorum, in contrast, employment grew faster than the number of production units (respectively 40.8% and 60%) in the same period.

These spectacular growth rates were mainly achieved in the flexibly organised, labour-intensive sectors. In particular, textile and clothing firms became the initiators of export orientated production. Production networks based on a long tradition of working together enabled textile firms to take off in these regions. In Denizli textile production[2], in Gaziantep textiles and machinery and, in Çorum non-metallic products (bricks etc.), machinery and textiles have been the leading sectors of growth. In sharp contrast with production in Turkey's regional centres and less developed regions, export orientated manufacturing with strong links to external markets is characteristic of production in these new manufacturing nodes (Eraydin 1998b). From Denizli 44.3%, from Gaziantep 36.8% and from Çorum 10.1% of total manufacturing production is exported to various countries according to the latest figures[3]. In some sectors, however, exports are far more prominent. For example, 61.9% of the textile products of Denizli and 25% cent of textile and chemicals and plastics products in Çorum are exported annually. The analysis presented in this chapter focuses on the conditions, networks and coalitions created by local economic actors and institutions that made this unexpected growth possible.

Take-off: The Successful Interplay of Policies, Institutions and Local Producers

Economic take-off in the Denizli, Gaziantep and Çorum industrial districts goes back to 1980s, and can be explained in terms of the interplay of liberal macro-economic policies, local capacities, and policies designed to foster local growth.

The 1980s were a turning point in Turkish economic policy. Earlier protectionism (see Kazgan 1985, Boratav 1988) was replaced by an increasing reliance on market forces. Initially, foreign trade and exchange controls were freed up, but further liberalisation in 1984 saw foreign exchange controls and quotas on imports dismantled and tariffs revised (Olgun and Togan 1984). Firms began to work in a more competitive environment. Liberalisation continued with export promotion policies and depreciation of the exchange rate (Senses 1989). Direct subsidies on exports reached 20% of production costs in the first half of the 1980s. New labour laws curtailed union power and the index of real wages fell dramatically from 100 in 1979 to 71 in 1982. This shift in factor prices gave labour intensive production cost advantages in export markets[4] (Eraydin and Erendil 1999). Clothing firms were the first to take advantage of these new conditions. These first firms were mainly located in the metropolitan areas, but several clusters of firms in the new industrial districts developed to capitalise on the new opportunities. They were able to respond quickly because of existing enterprise networks, local experience in handicrafts, and local knowledge generated in traditional labour intensive sectors (Eraydin 1998b).

In fact, in both Denizli and Gaziantep the history of the textile and clothing sector goes back centuries. In Denizli, spinning and weaving that had begun in the Buldan and Babadağ subcentres, surged in the 1960s with the adoption of electrically operated looms. A few state-owned firms[5], larger factories set up by cooperatives, and also 'worker-enterprises' were important in the 1960s and 1970s. These 'worker-enterprises', set up by people working abroad (especially in Germany), showed that businesses could be established on a collaborative basis, although they were of limited success because they were undercapitalised and lacked management expertise. Production rose in the 1960s as several private firms developed subcontacting relationships with local SMEs. In Gaziantep, textiles and transportation equipment production began to flourish in the 1950s. Gaziantep had benefited from Marshall Aid that had sponsored agricultural mechanisation and major highway construction. An opportunity arose for former handicraft shops to become vehicle repair workshops. In 1970, to support the technological improvement of these flourishing small-scale enterprises, a collaborative project between the Ministry of Trade and Industry and the United Nations set up a pilot industrial unit KÜSGEM (Small Scale Industry Development Centre). As part of this project, 200 hectares of land, with infrastructure, a central laboratory and testing areas, were provided for small industries. In the late 1970s, another central government institution, the Small and Medium Enterprises Development Centre (KOSGEB), helped SMEs to improve their technologies and to become more effective contributors to the economy. More new firms were created, but with only limited improvements in technological levels. However, the experience gained through these initiatives was the foundation of present day machinery production in Gaziantep. In Çorum, development was initiated somewhat differently (Eraydin 1998a). Initial development was focused on resource-based firms (engaged in flour milling and brick and tile making) using local resource advantages. More recently, machinery production has flourished as a spin-off from these resource-based activities.

Up to the 1980s, all these new industrial districts also benefited from special incentives and financial support programmes designed for less developed regions. These *encouragement incentives*[6] were provided by the government, and were similar to those available in other less developed regions in Turkey. However, they generated mainly locally orientated businesses that were unwilling to compete with metropolitan-based firms.

Through the 1980s, there was radical change in central government policies. First, central government adopted export-orientated development policies and offered export promotion incentives and special support for firms engaged in foreign trade. These incentives and support favoured small as well as large firms, with small firms being able to capitalise on their existing capacities and their labour cost advantages. Additionally, small firms were able to obtain capital equipment second-hand as large and medium size firms, who became eligible to import machinery without duty, either enlarged their production capacities or renewed their machinery stock. They sold their not-too-old machinery to small firms at reasonable prices (Eraydin 1995). Later, in the 1990s, these incentive mechanisms[7] were extended to small firms. Second, several credit institutions reformed themselves in order to support export orientated firms, including small production units. These included the People's Bank (Halk Bankası), the Turkish Development Bank (Türkiye Kalkınma Bankası) and Eximbank. They began to provide credit at below market rates, but only a limited number of small businesses were able to make use of these facilities[8]. In Denizli, at the initial stage, only 25.2% of large firms and 20.5% of small firms used these credit facilities. The ratios were even lower in Gaziantep (15.4% of large firms and 7.7% of small firms). They sought credit only after they had begun production. Third, new roles were defined by KOSGEB, emphasising information on technological improvement, laboratory services, quality and innovativeness. But, in Denizli, Gaziantep and Çorum, entrepreneurs were not well informed about these facilities. Fourth, the standard tools for industrial support – providing land with infrastructure in *Industrial Estates* for medium and large enterprises, and land, infrastructure and working places in *Small Industrial Districts* for small firms – were widely used. These industrial sites were built by the central government and paid for by entrepreneurs through low interest loans. On completion, the sites were operated by local councils comprising local government representatives and local entrepreneurs. There is a wide consensus among local actors that these facilities have been important in stimulating the growth of small and medium industrial enterprises[9].

Clearly, during the 1980s, the Turkish government realised the importance of small enterprises as a source of economic strength that goes beyond the creation of employment, for which they had long been recognised. Their production and export capacities had generally been underestimated but, during the 1980s, measures were designed to realise their export growth potential. There have been two results from these policies.

First, the successful performance of leading firms has drawn small enterprises to export orientated production. Various socially embedded practices and conventions have facilitated this process that can be defined as *'following successful examples'* (Erendil 1998). *'Compatriotism'*, similar to craft values, is one of these social proprieties. Being from the same area (village, town and city) has always been very important in social interactions in the parts of Turkey where Denizli, Gaziantep and Çorum are located. This not only helps collaborative networks to form, it also imposes certain responsibilities on network members, and provides new entrants with places in the newly emerging production networks. The networks provide different forms of support, ranging from information and technical knowledge up to financial aid from compatriots. Capital, support is particularly important when new firms are set up since their owners are often wary of capitalistic relations in production and banking environments. Small entrepreneurs usually use their own capital and get support from their family and relatives, but if their financial resources are not adequate they depend upon their fellow citizens. This dependence fades with success. In later stages of growth, compatriot relations link small enterprises into production networks. Although market mechanisms undermine these peasant-culture values, they remain an important social institution. For example, in the Denizli clothing industry it is possible to still identify production networks initiated by entrepreneurs from Babadağ (a small town in the region), which had been vital in building relationships of solidarity and trust among the most prominent entrepreneurs.

Second, the number of small firms has increased dramatically through processes of spin-off. In the absence of financial constraints, spin-offs from parent firms have been set up with family finance. Growth tends to put strain on this cooperative form of ownership, and when it breaks some members take their own shares of capital to form their own businesses. However, even though ownership may change, collaboration continues. Usually the spin-off enterprises protect their linkages with their parent firms and become production network partners, competing with one another but also collaborating to fulfil large orders.

The Consequences of Growth: Fragmented Collaboration and Institutional Split

Following the sudden growth stimulated by local initiatives, small firms in Denizli, Gaziantep and Çorum began to face difficulties in the late 1980s. Production quality lagged behind international market requirements, technology needed updating, collaborative networks began to break down as firms started to follow different strategies, and production began to fragment.

Fragmented Strategies

At this time, larger firms began to internalise elements of production they had previously subcontracted because the small subcontractors could not meet the required quality standards. This strategy was particularly prevalent among textile and clothing firms producing high value added goods for the high end of the European market. It did not end subcontracting, but put it on a more flexible and selective basis.

Small enterprises with competitive power formed several cooperative institutions to allow them to compete with the networks dominated by the leading larger firms. They formed coordinating and service units, such as the Aegean Ready-Garment Producers Association (EGS). EGS was created in 1993, by 464 small producers, 60% of whom were from Denizli. In the first instance, they tried to form an association to discuss the problems they faced in manufacturing for export. While the member firms stayed autonomous, they first formed an export company and then several support companies: a service firm to procure inputs (EGESER); a transportation firm (EGSNAK); and an insurance company (EGS Sigorta). In 1995, they also founded a bank, the EGS Bank. The services provided by the firms in the EGS Group were very influential in promoting the members' exports, and, in 1996 and 1997, the EGS Foreign Trade company became the leading export firm in Turkey. Recently, the EGS Group also began to open retail stores abroad, especially in Europe, to sell the products of their members under an original brand name.

This model of collaboration is somewhat different to the 'agrupamientes industrialer' in Guadalajara, where the groups are led by a leader defined by the entrepreneurs' association (Rabelotti 1997). In fact, EGS, as a region-based institution that represents the common interests of similar firms, has been more successful than many export-promoting organisations – foreign trade companies (FTCs), associations of small exporters and sectorally-based semi-public exporter associations, for example (Eraydin 1993).

Small, low technology firms in the textile and clothing sector constitute a separate segment, making low quality products for the domestic, East European and Russian markets. Local mediators played an important role in promoting the exports of these firms who had little experience of the market. They coordinated production among small firms and supplied raw materials, an arrangement similar to the classical putting-out system. In this respect, the activities of mediators in Denizli are similar to arrangements that have been identified in developing countries, such as in Trippur in India (Cawthorne 1995) and in the Sinos Valley in Brazil (Schmitz 1995) (see Erendil 1998).

These different strategies of cooperation and mediator-led production supplemented rather than replaced subcontracting. A number of studies have examined the extensive subcontracting relationships and the intricate internal dynamics of production in Denizli, Gaziantep and Çorum (Çınar, Evcimen and Kaytaz 1988, Eraydin and Erendil 1999, Kayasu 1995). Still, the labour intensive stages of production are mostly subcontracted to small specialist firms (Eraydin 1997, Kaytaz 1994) or to homeworkers (Çinar 1989, Kümbetoglu 1996, Lordoglu 1990, White 1994). The concentration of specialised activities and inter-firm interaction has become the main advantages of these clusters. As a result, each segment of production has benefited from locally embedded interaction systems and inter-organisational learning processes to improve their competitiveness.

Exploitation of Locally Embedded Relations

In the Gaziantep and Çorum industrial districts, survey results show that learning-by-interaction between producers and customers is a significant source of competitive advantage for half of all local business where tacit knowledge is important but technological change is relatively slow (Eraydin 1998a,b). Machinery producers claimed that they gained technical information, new designs and technologies from their customers, in addition to information on market conditions. In Çorum, these close relationships are between firms in the resource-based and machinery sectors, with machinery makers either customising or specially designing equipment for the clients in the resource-based sector. In 1999, these firms supplied nearly a quarter of local machinery demand: a greater proportion in flourmills, animal feed producing factories and metal products and machinery production units[10], but a relatively lower proportion in the newer fields of production.

Local machinery producers are also essential as part of a process of imitation in which exogenous codified knowledge can be blended with local tacit knowledge to generate local solutions. As one entrepreneur remarked, '… a design can be developed by the customer or can be taken from abroad and imitated by local firms at a lower cost'. For this reason, clients are eager to transfer information acquired at fairs or during visits to foreign producers, and remain reluctant to buy new product innovations off-the-shelf, nationally or internationally.

In Denizli, Gaziantep and Çorum attempts to improve, transfer and adjust technologies have been important. Using either new machines or their own capacities, 44.9% of firms in these districts have been shown to be technologically innovative. In Gaziantep, out of 320 firms, 45 have product patents and 10 have patents related to production processes. In Denizli, 60 out of 416 firms (14%) are engaged in innovative research, a far greater proportion than in Turkey as a whole (Eraydin 1998b). Facilitating this process in SMEs, engineers, entrepreneurs and skilled workers have become integrated into production activities. In addition, locally embedded relations have contributed significantly to the creation of local innovative attitudes. However, as in the well-known example of Emilia-Romagna, collaboration between firms and universities tends to be weak. Only when firms can not solve their problems themselves do they attempt to contact universities or public research institutions[11].

The development experience of these three industrial districts also exemplifies the importance and merits of leadership. Leader firms are important in generating growth motivation and the imitation of best practice is important for initiating growth. Many local entrepreneurs claimed that concrete evidence has been the most important encouragement for them. In fact, 'they believe what they see'. They are less interested in experimentation, and imitation is not a particularly good strategy for rapid knowledge accumulation (Maskell and Malmberg 1999a).

Institutional Fragmentation

At the beginning of the industrialisation phase, international competition and the need to change drew firms together in associations and institutions. However, these associations and institutions tended to have different values and goals, resulting in power struggles and conflicts. One group of associations has craft values enshrined in a communitarian ideology of cooperation and consensus and strong ties among fellow citizens in nearby localities. The small entrepreneurs who belong to this group tend to organise in artisan associations or cooperatives.

In contrast, relatively larger enterprises usually prefer to form separate institutions with distinctive sets of values. The Chamber of Industry is one of these associations formed by larger enterprises, but local Chambers of Industry tend to be conservative and appeal to only one section of business. Larger firms, who are eager to engage with international markets and take risks, prefer to organise under the local Associations of Industrial Entrepreneurs (SIADs), following the example of the largest entrepreneurs in Turkey (TUSIAD). They are said to favour globalisation and export orientated development. But recently, even the SIADs (ÇORUMSIAD, DESIAD, etc.) have been said to lack dynamics and aggressive export polices. Now, Young Entrepreneurs Associations have assumed that mantle.

Still other groups draw together business people with different kinds of economic understandings and different social attitudes. The most important of these groups is the Independent Entrepreneurs Association (MUSIAD). Although the formal name includes the word 'independent', everyone knows that the 'M' denotes 'Muslim'. The association was formed by Muslim businessmen to counter a hostile socio-political environment, and to dispel misconceptions about the incompatibility of Islam with entrepreneurial activity (Buğra, 1999). In fact, MUSIAD can be seen as a form of class organisation against the association of large entrepreneurs. They use religion as a bond but downplay it for fear of being seen as part of a narrow set of Islamic interests or fundamentalist groups.

Institutions were also set up to appropriate government benefits (from both central and local government) and to gain the ear of government. The Exporters Association was set up for just this reason. However, the proliferation and fragmentation of associations and institutions has been strongly motivated by the desire to have easier access to government and to pursue sectional goals. This is hardly surprising when government support has been defined as the basic solution to secure growth and deal with the problems of growth. However, the patriarchal relations between firms in some associations, in which lead firms support and guide others, but not to the extent that they can overtake them, has also created internal power struggles within associations and has stifled individual creative strategies.

Crisis: Traditional Solutions for New Problems

Interactive relations and collective learning were only able to induce minor product diversification in Denizli, Gaziantep and Çorum. The initial growth of exports from Denizli was of towels and bathrobes[12], to which ready-made garments were later added (Pınarcıoğlu 1998). Diversification in Gaziantep, in the 1990s introduced new textile products and garment production, while in Çorum, the experience gained in machinery production field initiated diversified high value production in this sector.

All the changes were partial improvement strategies constrained by fear of the unknown. In this period of growth and change, the owners of the lead firms recognised that production needed reorganising, and that management and marketing were as important as production. As a result, because they were unwilling to become dependent on professionals, they sent their daughters and sons for management training, especially in US universities (Saraçoğlu, 1993). During the 1990s, this second generation became managers. They wanted to adopt more aggressive strategies, but were constrained by strong patriarchal, risk-averse, family relationships, even though the consequences of this conservatism were recognised. Local and national authorities were unable to affect the situation and continued to confine themselves to solving the problems of existing production units.

In the mid 1990s, it became more obvious that existing local networks were not capable of updating the technological competence of firms and to revive their competitive power, although the industrial districts studied have been successful in terms of increasing numbers of firms and employees. Intensive local interaction brought inbreeding and led to growth without radical change in existing products or production practices, weakening firms' positions in national and international markets. The macro economic problems of the years 1997 to 1999, that had their origins in Asia and Russia, exacerbated the problems in Turkey and the regions of Denizli, Gaziantep and Çorum. In 1999, national GNP fell by 6.4%, and industrial production fell by 3.9% largely as a result of problems in the Russian economy.

These economic problems created unrest, institutional conflict, and fragmentation. The economic downturn caused the fragmentation of groups, and undermined cooperative action based on trust and reciprocity. As a result, collaborative social relations involving trust and mutual support have been eroded. New firms have been particularly vulnerable (Pınarcıoğlu 1998, Erendil 1998, Özcan 1995) because they lack the experience to deal with crisis. Small firms have been badly affected. According to local authorities, at least 50% of small firms have cut production, and some have stopped production altogether. It has taken some time for business people to overcome the crisis, to address the sources of their problems, and to develop strategies. In effect, they have had to unlearn the successful habits of the past that might hinder future success (Maskell and Malmberg 1999a). In fact, some producers claimed that this crisis period has been 'the best training programme'.

To deal with the new conditions was obviously quite difficult for the industrial districts that were specialised in labour-intensive or craft-type products (Eraydin 1997). Some regions have successfully transformed themselves by becoming innovative learning regions (Cooke 1996), a strategy denied to some districts because of asset erosion or lock-in (Maskell and Malmberg 1999a). The reluctance of local actors to accept radical changes can be added to this list (Harrison 1994b). The main economic growth achievements of many regions are mainly founded on path dependent learning and adaptation, but, at the same time, they lack the power and the capacity to follow new path-breaking courses (Amin 1999).

There is limited experience of the courses of action local production systems might adopt to cope with change, but from the existing literature five broad pointers can be identified. First, there is Scott's (1994) suggestion of 'collective activism'. For Locke (1995), this involved the development of associations and interest groups to mediate industrial conflict and diffuse information. For Ottai (1996), it necessitated more conscious and organised relationships between enterprises, and for Storper (1993), it involved moving up the price-quality curve.

Second, necessary restructuring has been seen to involve external linkages to make sure that a 'milieu' or district does not stew in its own juice (Johannessen, Dova and Olsen 1997). Increasingly, these necessary ties are seen to be global (Amin 1999), and merging with a globally competitive firm is a possible strategy, but one that has stimulated limited interest in the business environment (Cooke 1996). Global firms, it seems, prefer to increase linkages with firms in industrial districts without any ownership commitment.

Third, there is the possibility that 'lead firms' in industrial districts might generate new technologies to stimulate local regeneration. But, though they might succeed themselves, they seem to be unable to inject strategic capability into the clusters within which they are embedded. As the Benetton case shows, some of these firms might even become multinational (Harrison 1994b), which then changes their ties with their local production environment.

Fourth, local partners might implant new technology in a region. The literature emphasises local cooperation and a new division of labour between social partners to realise the benefits from implanted new technology (Amin 1999). To succeed in this way, there is a need for policy action from below and through democratic dialogue. Local development coalitions (Schmitz and Musyck 1994) are important in this respect, but as emphasised earlier, the state is still seen as an eligible mediator (Schmitz 1998, Scott 1994).

Fifth, local institutional change has been indicated as necessary before new economic priorities can be pursued in an industrial district. Depending on the priorities set, a range of policy options is available, from locally identified policy-based programmes and incubator project solutions (Amin 1999) to 'place marketing' programmes.

In fact, in each of the Turkish industrial districts studied here – Denizli, Gaziantep and Çorum – one or more of these policy options has been adopted to address local crisis conditions. In Denizli, most of the firms, especially the ones that produced for the domestic market, had been deeply affected by the late 1990s recession. In this period, 11,000 registered workers lost their jobs, prompting urgent action. The policy option selected was to strengthen local competitive capacities through cost reduction projects. Two projects were particularly noteworthy; a new Free Trade Zone, which will provide tax advantages to export firms within it, and the 'Transformation 2000' project, which aimed to upgrade the technological basis of firms through an active information system among EGS partners.

In Çorum, the major emphasis was on place-marketing and the search for new partners and fields of production. Advertising through the mass media was used to create a new image of the district. Entrepreneurs' associations raised its profile, especially the Çorum Industrial Entrepreneurs Association, inviting institutions like the World Bank, the World Academy of Local Development and the State Planning Organisation of Turkey to the area. In addition, a group of entrepreneurs formed a committee to propose to the Department of National Defense Industry that they were able to produce parts for military equipment.

Gaziantep was among the first districts to face the economic difficulties of the 1990s, when its Middle Eastern markets, especially in Iraq, disappeared in the 1991 Gulf War. The district's search for new markets has been hindered by macro economic difficulties and inadequate export credit. They have managed to set up a Business Development Centre with the support of the European Union and they are trying to set up an information network not only in the district but also across the less developed parts of southeast Anatolia.

There can be no certainty that the policy initiatives that have been adopted in the three study districts will succeed. It has been the smallest firms that have been most seriously affected, either ceasing production or continuing to produce without recovering their costs for social reasons. The larger firms have wider options and, for example, seek work as subcontractors to foreign firms. In practice, however, they have adopted traditional cost minimisation strategies in the same way as firms in Italian industrial districts (Cooke and Morgan 1994).

Conclusions

In recent years, regional development theories focused on large manufacturing units, infrastructure provision and public intervention have failed both to explain regional dynamics and to direct development processes. In their place, an interpretation of local growth that recognises the social construction of local economies has developed in a literature on industrial districts, clustering, regional innovation systems, learning regions and innovative milieus. The emphasis in this new generation of thinking is on embeddedness and the incorporation of firms into local networks of reciprocal relationships based on trust and cooperation. Experience in industrial districts suggests that these locally embedded relationships can create local competitive advantage in a globalised economic environment, although the level of success has varied from place to place. A growing literature now questions the universal applicability of this view of local growth mechanisms based on embeddedness, suggesting that it might be unable to sustain competitive advantages and stimulate innovation (see, for example, Staber 1997, Cooke and Morgan 1994). The experience of growth and crisis in the three new industrial clusters of Denizli, Gaziantep and Çorum in Turkey sheds light on these issues.

From the analysis in this chapter of industrial districts in Turkey, it is clear that locally embedded relations are important in initiating and sustaining local growth and competitive power, but that the benefits have tended to be inadequate and temporary. Although the local social milieu enabled the exchange of information and continuous learning about techniques and opportunities, experience shows that there is always the danger that the information and knowledge that is created is not internationally competitive. The transition from imitation to learning and innovation needs specific skills and a new synthesis of tacit and codified knowledge. This has not happened in the Turkish districts, and it is very difficult to claim that many industrial clusters, in even the advanced part of the world, have these capacities. The limited learning capacity of regional actors can be a serious economic weakness, and though learning may appear to be occurring, it is in periods of crisis that the real situation is seen. The experience in Turkey would suggest that learning was more apparent than real, and that no culture of experimentation had developed. So, when faced with crisis conditions, traditional cost minimisation strategies were employed just as in many other parts of the world. The experience in Denizli shows, nevertheless, that small firms can act as buffer in difficult economic times and that large firms can survive at the expense of these smaller, weaker firms.

Government policies and continuing support through incentives may have initiated local growth and the establishment of local networks and collaborative relationships in these Turkish districts. But, plainly, it has been insufficient to create lasting and self-sustaining structures that might form the basis of a 'learning region' or a 'regional innovation system'. In addition, local 'institutional thickness', though extensive, has also proved insufficient to support self-sustaining growth and counter externally imposed crisis conditions. It is possible that fragmentation of this institutional support has undermined its effectiveness in Turkey. The conditions of growth and crisis in Turkey show very clearly that the strength and resilience of industrial districts can not easily be judged from a count of characteristics that might be necessary but not sufficient to promote and sustain local economic growth. The analysis suggests that we need much more fine-grained information on the workings of firms within networks and their ties to place to understand their resilience and success.

Notes

1   As a result of this rapid growth, this district's share of total manufacturing firms rose from 1.61% to 1.74% and its share of employees rose from 0.91% to 1.49% between 1990 and 1995.
2   In 1996, the share of textile and clothing firms among all manufacturing firms was 65.6%, and the share of these firms in total manufacturing employment rose to 76.3%.
3   Larger firms dominate exports (69% of the total).
4   The new institutions and financial incentives of the 1980s led many firms to export. Industrial exports rose spectacularly from US$ 1,401 million in 1980 to US$ 12,959 million in 1990 and US$ 21,637 million in 1995 (at current prices) primarily as a result of improved competitiveness, export incentives and subdued domestic demand (Kazgan 1985, Senses 1989). Similarly, clothing exports, which amounted to US$ 131 in 1980, reached US$ 1,208 in 1985 and US$ 6,121 million dollars in 1995, some 32% of the total manufacturing exports.
5   Denizli-Sümerbank Spinning Mill set up in 1953, added new weaving units in 1964; Gaziantep Alcoholic Drinks factory was set up in 1942 and a cement factory that was completed in 1961.
6   The incentives are mainly sector specific, and except in some regions (such as inner parts of Metropolitan Areas) all firms can get sector specific incentives. However, firms in less developed regions (with priority for development) get additional incentives. The 'Resource Utilisation Premium' that reached 30-40% of total fixed investment increased the attractiveness of these areas after the 1980s. Gaziantep received special grants for less developed regions, tax exemptions and low interest credits in the 1968-1972 and 1978-1980 periods, while Denizli was eligible for special grants in the 1973-1981 period, and Çorum from 1973 onwards (Sarica 1991).
7   In Denizli 54.3% of the firms that employ more than 10 persons have made use of incentives. Textile and clothing firms had greater access to incentive schemes (62.8%), while the proportion of all firms and textile units that benefited from the incentive schemes was lower (39.2% of textile firms) in Gaziantep and Çorum.
8   In advanced countries there are financial institutions that help small firms obtain capital. In Turkey this service is provided by Credit Guarantee Fund (KGF). KGF, as an intermediary institution that has been formed by various institutions related to small enterprises, provides deposits or guarantees for small firms that employ less than 200 workers. However, most SMEs do not know of this facility and are unable to use it.

9   The Industrial Estate in Denizli is relatively new. The location of the industrial estate was decided in the 1975 Development Plan of Denizli. However, construction did not begin until 1985 owing to financial difficulties. The infrastructure was completed in 1988. In 1998, there were 123 factories operating on this estate, besides the 15 rented factory buildings. Most of the factories in the industrial estate are textile firms (82 firms) engaged in weaving, dying and printing, but there are also firms that produce metal goods, plastics and non-metallic products (Erendil 1998). After the allocation of 139 land parcels available to the firms, the development of a second industrial estate has been initiated in the Çardak area with an additional 100 new parcels of land.

10   The share of locally produced machinery is low in brick and tile factories (8.6%), whereas the share of local machinery is relatively higher in textile and clothing firms (20%). The packaging industry, chemical and plastic goods production and basic mineral industry do not use local machinery. They are insufficiently developed to have a spin-off effect.

11   Eight firms out of 69 worked with public research institutes to acquire new technology. Among these institutions are the Istanbul Technical University, the Department of Wood Works in Hacettepe University, and Departments of Agriculture of several universities, and Marmara University for soil analysis. The firms also have contacts with the Centre for Small and Medium Enterprises, of the Ministry of Industry, and Institute of Search of Mines (MTA).

12   These two products made up 80% of all exports in the 1990-94 period.

References

Amin, A. and Thrift, N. (1994), 'Living in the Global', in A. Amin and N. Thrift (eds), *Globalization, Institutions and Regional Development in Europe*, Oxford University Press, Oxford, pp. 1-22.

Amin, A. and Thrift, N. (1995), 'Globalisation, institutional thickness and the local economy', in P. Healey, S. Cameron, S. Davoudi, S. Graham and A. Madani-Pour (eds), *Managing Cities: The New Urban Context*, John Wiley, London, pp. 91-108.

Amin, A. (1999), 'The Emilian model: Institutional challenges', *European Planning Studies*, vol. 7, no. 4, pp. 389-405.

Amin, A. and Cohendet, P. (1999), 'Learning and adaptation in decentralised business networks', *Environment and Planning D: Society and Space*, vol. 17, pp. 87-104.

Angel, D. (1991), 'High-technology agglomeration and the labor market: The case of Silicon Valley', *Environment and Planning A*, vol. 23, no. 10, pp. 1501-1516.

Asheim, B.T. (1996), 'Industrial districts as "learning regions": A condition for prosperity', *European Planning Studies*, vol. 4, no. 4, pp. 379-397.

Becattini G. (1991), 'The industrial district as a creative milieu', in G. Benko and M. Dunford (eds), *Industrial Change and Regional Development*, Belhaven, London, pp. 102-113.

Belussi, F. (1996), 'Local systems, industrial districts and institutional networks: Towards a new evolutionary paradigm of industrial economics', *European Planning Studies*, vol. 4, pp. 5-26.

Boratav, K. (1988), *Turkiye Iktisat Tarihi*, Istanbul, Gercek.

Brusco, S. (1986), 'Small firms and industrial districts: The experience of Italy', in D. Keeble and E. Wever (eds), *New Firms and Regional Development*, Croom Helm, London, pp. 184-202.

Buğra, A. (1999), *Islam In Economic Organizations*, Friedrich Ebert Yayınları, Istanbul.

Camagni, R. (1991), 'Local milieu, uncertainty and innovation networks: Towards a new dynamic theory of economic space', in R. Camagni (ed.), *Innovation Networks*, Belhaven, London, pp. 121-144.

Cawthorne, P.M. (1995), 'The rise of networks and markets of a South Indian town: The example of Tiruppur's cotton knitwear industry', *World Development*, vol. 23, pp. 43-56.

Chronaki, Z., Hadjimichalis,L., Labriandis, L. and Vaiou, D. (1993), 'Diffused industrialisation in Thassaloniki: From expansion to crisis', *International Journal of Urban and Regional Research*, vol. 17, no. 3, pp. 178-193.

Cinar, M., Evcimen, G. and Kaytaz, M. (1988), 'The present day status of small-scale, industries (sanatkar) in Bursa, Turkey', *International Journal of Middle East Studies*, vol. 20, pp. 287-301.

Cinar, M. (1989), *Taking Work at Home: Disguised Female Employment in Urban Turkey*, Working Paper No. 8810, Loyola University of Chicago, School of Business Administration, Chicago.

Cooke P. (1996), 'Building a twenty-first century regional economy in Emilia-Romagna', *European Planning Studies*, vol. 4, no. 1, pp. 53-62.

Cooke, P, Uranga, M. G. and Etxebarria, G. (1998), 'Regional systems of innovation: An evolutionary perspective', *Environment and Planning A*, vol. 30, pp. 1563-1584.

Cooke, P. and Morgan, K. (1998), *The Associational Economy: Firms, Regions and Innovation*, Oxford University Press, Oxford.

Cooke, P. (1998), 'Introduction: Origins of the concept', in H-J. Brazyk, P. Cooke and M. Heidenreich (eds), *Regional Innovation Systems: The Role of Governance in a Globalized World*, UCL Press, London, pp. 2-27.

Cooke, P. and Morgan, K. (1994), 'Growth regions under duress: Renewal strategies in Baden-Württemberg and Emilia Romagna', in A. Amin and N. Thrift (eds), *Globalisation, Institutions and Regional Development in Europe*, Oxford University Press, Oxford, pp. 91-117.

Coriat, B. and Bianchi, R. (1995), 'A European response to the Japanese challenge', in B. Andersen *et al.* (eds), *Europe's Next Step*, Frank Cass, Ilford, pp. 59-77.

Eraydin, A. (1999), Türkiyedeki Sanayi Gelişmesinin Anadolu'ya Yaygınlaşması Ve Son Dönemde Gelişen Yeni Sanayi Odakları', *Çarktan Çip'e*, Tarih Vakfı, Istanbul, pp. 257-278.

Eraydin, A. And Erendil, A. (1999), *Dis Pazarlara Acilan Konfeksiyon Sanayinde Yeni Uretim Surecleri Ve Kadin Isgucunun Bu Surece Katilim Bicimleri*, KKSGM, Ankara.

Eraydin, A. (1993), 'Business behavior and restructuring of the Turkish economy', in C. Rogerson, E. Schamp and G.J.R. Linge (eds), *Finance, Institutions and Industrial Change: Spatial Perspectives*, Gruyter, Berlin, pp. 183-203.

Eraydin, A. (1998a), *From an Underdeveloped Region to a Locality: The Experience of Çorum*, unpublished Paper prepared for the World Bank.

Eraydin, A. (1998b), *The Role of Regulation Mechanisms and Public Policies at the Emergence of the New Industrial Districts*, Paper presented at the symposium on 'New Nodes of Growth in Turkey: Gaziantep and Denizli', Ankara.

Eraydin, A. (1997), 'LDC industrial districts: The challenge of the periphery', in K.I. Westeren (ed.), *Cross Border Cooperation and Strategies for Development in Peripheral Regions*, Nord-Trondelags, Forskning, Oslo, pp. 411-436.

Eraydin, A. and Erendil, A. (1999), 'The role of female labour in industrial restructuring: New production processes and labour market relations in the Istanbul clothing Industry', *Gender, Place And Culture*, vol. 6, no. 3, pp. 259-272.

Erendil, A. (1998), *Using Critical Realist Approach in Geographical Research: An Attempt to Analyze the Transforming Nature of Production and Reproduction in Denizli*, unpublished PhD thesis, Department of City and Regional Planning, Middle East Technical University, Ankara.

Glasmeier, A. (1991), 'Technological discontinuities and flexible production networks: The case of Switzerland and the world watch industry', *Research Policy*, vol. 20, pp. 469-485.

Glasmeier, A. (1994), 'Flexible districts, flexible regions? The institutional and cultural limits to districts in an era of globalisation and technological paradigm shift', in A. Amin and N. Thrift (eds), *Globalisation, Institutions and Regional Development in Europe*, Oxford University Press, Oxford, pp. 118-146.

Grabher, G. (1993), 'Rediscovering the social in the economics of interfirm relations', in G. Grabher (ed.), *The Embedded Firm. On Socioeconomics of Industrial Networks*, Routledge, London, pp. 1-33.

Granovetter, M. (1985), 'Economic action and social structure: The problem of embeddedness', *American Journal of Sociology*, vol. 19, pp. 418-510.

Gregersen, B. and Johnson, B. (1997), 'Learning economies, innovation systems and European integration', *Regional Studies*, vol. 31, no. 5, pp. 479-490.

Harrison, B. (1994a), 'The Italian industrial districts and the crisis of cooperative form: Part I', *European Planning Studies*, vol. 2, no. 1, pp. 3-22.

Harrison, B. (1994b), 'The italian industrial districts and the crisis of cooperative form: Part II', *European Planning Studies*, vol 2, no. 2, pp. 159-174.

Heidenreich, M. and Krauss, G. (1998), 'The Baden-Württemberg production and innovation regime: Past success and new challenges', in H-J. Brazyk, P. Cooke and M. Heidenreich (eds), *Regional Innovation Systems: The Role of Governance in a Globalized World*, UCL Press, London, pp. 214-244.

Johannessen, J.A., Dolva, J.O. and B. Olsen, (1997), 'Organising innovation: Integrating knowledge systems', *European Planning Studies*, vol. l5, no. 3, pp. 331-349.

Kayasu, S. (1995), *Local Production Organization Oriented Towards Global Markets: Subcontracting Relationships in the Clothing Industry*, unpublished PhD thesis, Department of City and Regional Planning, Middle East Technical University, Ankara.

Kaytaz, M. (1994), 'Subcontracting practice in the Turkish textile and metal working industries', in F. Senses (ed.), *Recent Industrialization Experience of Turkey in a Global Context*, Greenwood Press, London, pp. 141-154.

Kazgan, G. (1985), *Ekonomide Disa Acik Buyume*, Altin Kitaplar, Istanbul.

Kumbetoglu, B. (1996), 'Gizli Isciler: Kadinlar Ve Bir Alan Arastirmasi', in S. Cakir and N. Akgokce (eds), *Kadin Arastirmalarinda Yontem*, Sel Yayincilik, Ankara, pp. 230-238.

Locke, R. (1995), *Remaking the Italian Economy*, Cornell University Press, Ithaca and London.

Lordoglu, K. (1990), *Eve Is Verme Sistemi Icinde Kadin Isgucu Uzerine Bir Alan Arastirmasi*, Friedrich Ebert Vakfi Yayınları, Istanbul.

Malmberg, A. and Maskell, P. (1997), 'Towards an explanation of regional specialisation and industry agglomeration', *European Planning Studies*, vol. 5, no. 1, pp. 25-41.

Malmberg, A. (1996), 'Industrial geography: Agglomeration and local milieu', *Progress in Human Geography*, vol. 20, no. 3, pp. 392-403.

Maskell, P. and Malmberg, A. (1999a), 'Localised learning and industrial competitiveness', *Cambridge Journal of Economics*, vol. 23, pp. 167-185.

Maskell, P. and Malmberg, A. (1999b), 'The competitiveness of firms and regions: Ubiquitification and the importance of localized learning', *European Urban and Regional Studies*, vol. 1, pp. 9-25.

Morgan, K. (1997), 'The learning region: Institutions, innovation and regional renewal, *Regional Studies*, vol. 31, no. 5, pp. 491-503.

Olgun, H. and Togan, S. (1984), *Turk Ekonomisinin Dunya Ekonomisine Entegrasyonu*, Enka, Istanbul.

Ottai, G.D. (1996), 'Economic changes in the industrial district of Prato in the 1980s: Towards a more conscious and organized industrial district', *European Planning Studies*, vol. 4, no. 1, pp. 307-330.

Özcan, G.B. (1995), *Small Firms and Local Economic Development*, Avebury, Aldershot.

Pinarcioğlu, M. (1998), *Industrial Development and Local Change: The Rise of Textiles and Clothing Since the 1980s and Transformation in The Local Economies Of Bursa and Denizli*, unpublished PhD thesis, University College London.

Pyke F. and Senberger, W. (1991), *Industrial Districts and Local Economic Regeneration*, International Institute for Labour Studies, Geneva.

Rabelotti, R. (1997), *External Economies and Cooperation in Industrial Districts: A Comparison of Italy and Mexico*, Macmillan, Basingstoke.

Raco, M. (1999), 'Competition, collaboration and the new industrial districts: Examining the industrial turn in local economic development', *Urban Studies*, vol. 36 no. 5-6, pp. 951-968.

Saraçoğlu, Y. (1993), *Local Production Networks: An Opportunity For Development*, unpublished MCP thesis, Department of City and Regional Planning Middle East Technical University, Ankara.

Schmitz, H. (1995), 'Small shoemakers and fordist giants: The tale of a supercluster', *World Development*, vol. 23, pp. 9-28.

Schmitz, H. and Musyck, B. (1994), 'Industrial districts in Europe – policy lessons for developing-countries', *World Development*, vol. 22, no. 6, pp. 889-910.

Schmitz, H. (1998), *Responding to Global Competitive Pressure: Local Co-operation and Upgrading in the Sinos Valley*, IDS Working Paper 82, Institute of Development Studies, University of Sussex.

Scott, A.J. (1994), 'Variations on the theme of agglomeration and growth: The gem and the jewellery industry in Los Angeles and Bangkok', *Geoforum*, vol. 25, no. 3, pp. 249-263.

Senses, F. (1989), *1980 Sonrasi Ekonomi Politikalari Isiginda Turkiye'de Sanayilesme*, Yayinlari, Ankara.

Staber, U. (1996), 'Accounting for variations in the performance of industrial districts: The case of Baden-Württemberg', *International Journal of Urban and Regional Research*, vol. 20, pp. 299-316.

Staber, U. (1997), 'Specialization in a declining industrial district, *Growth and Change*, vol. 28, pp. 475-495.

Storper, M. (1993), 'Regional worlds of production: Learning and innovation in the technology districts of France, Italy and the USA', *Regional Studies*, vol. 27, pp. 433-455.

Storper, M. (1995), 'The resurgence of the regional economies. Ten years later: The region as a nexus of untraded interdependencies', *European Urban and Regional Studies*, vol. 2, no. 3, pp. 191-215.

Taylor, M. (1999) *Enterprise, Embeddedness and Exclusion: Buyer-Supplier Relations in a Small Developing Country*, Paper presented to the meeting of the IGU Commission on the Organisation of Industrial Space, Haifa and Beer Sheva, June.

Tödling, F. (1994), 'The uneven landscape of innovation poles: local embeddedness and global networks', in A. Amin and N. Thrift (eds), *Globalisation, Institutions and Regional Development in Europe*, Oxford University Press, Oxford, pp. 68-90.

White, J.B. (1994), *Money Makes us Relatives: Women's Labor in Urban Turkey*, University of Texas Press, Austin.

Chapter 16

Understanding Embeddedness

Michael Taylor and Simon Leonard

Introduction

The embeddedness model of local economic growth currently finds expression in a diverse range of literatures across the social sciences including work on clusters, new industrial spaces, regional innovation systems, economic and innovative milieu, and learning regions. It is an explanation of growth that emphasises the social construction of economies and the role of trust, reciprocal ties and loyalty in sustaining economic and commercial activity, and generating competitive power, productivity and profits. At its heart are social ties, fostered by proximity, that are reckoned to create social capital. And, social capital is now considered a vital complement to both money capital and human capital as a driver of place-based economic growth. Its significance derives from the mobilisation of knowledge and the processes of learning and innovation it is said to promote. Indeed, social capital is seen as a major driver of the 'knowledge economy' that seemingly all governments are striving to foster.

It is hardly surprising, therefore, that the embeddedness model of local growth is one of the principal planks in the policy platforms of both developed and developing countries and also the international agencies that attempt to foster economic growth more widely. It has an obvious appeal in an increasingly neo-liberal world. Growth is endogenous and requires local mobilisation. Social collaboration, as the generator of social capital, holds the potential to generate human capital and money capital. It also has the potential to foster learning, innovation and the knowledge economy, bringing international competitiveness in its wake. By this logic, governments and international agencies can act only as economic promoters and facilitators. They are certainly not subsidisers and the funders of either national or global Keynesianism. This is the logic that underpins World Bank policies to build social capital in the under-developed world, just as it is the logic of Blairite cluster policies of regional development in the UK, and the widely accepted and applied policies to foster the 'knowledge economy'. But, the empirical foundations on which this model is built are slight, and the question remains as to whether it is anything more than rhetoric in search of a reality.

The analyses of the embedded enterprise model drawn together in this volume have addressed its postulated workings in a range of empirical circumstances: in developed and developing countries; in primary production, secondary manufacturing and service industry sectors; among minority business enterprises, transnational corporations (TNCs) and small firms; in labour markets; and under different governance regimes. In all these circumstances there are fragments of support for the workings of the model, but nothing that could, by any stretch of the imagination, be called consistent or comprehensive. The combinations of supporting elements differ from place to place, are apparently fragile in the extreme, and are strongly time-specific.

Perhaps the strongest support for the embeddedness model of local economic growth comes from the reported research on financial and business services in the UK (*Search and Taylor*) and Australia (*Agnes*). In these case studies, trust, reciprocity and loyalty in network relationships were vital in the 1990s for the performance of business in sectors as diverse as foreign exchange dealing, currency swaps, master custody, and basic accountancy and legal services. They facilitated the local place-based transactions of large corporations and TNCs, and were an essential element in processes of new firm formation, mobilising the resources of business angels to assemble the temporary coalitions of new venturers to create businesses. At the same time, however, the nature and extent of local 'embeddedness' was quite distinctive, and essentially unique to each sector and to each place. What is more, that embeddedness was also suggested as being immensely fragile: vulnerable to the pressures generated by evolving information and communications technologies; affected by tendencies towards codification of financial service operations; susceptible to shifting fashions in corporate financing; and threatened by the impersonal, price-driven relationships that large companies can all too frequently impose on small firm networks. In short, embedded inter-firm relationships in financial and business services were jeopardised by the dead hand of corporatism.

The issue of 'proximity' that lies at the heart of the embeddedness model was also supported in a number of the case studies in the volume. As a foundation for the promotion of local economic growth, 'proximity' was recognised as playing an important role in recent Spanish experience (*Pallares-Barbera*) and in the functioning of non-metropolitan business services in the UK (*Search and Taylor*). However, while the study of the UK electronics industry (*Openshaw and Taylor*) also supports this interpretation, the support it gives is strongly time dependent, with the strongest embedded local business relationships having persisted only when government policy continued to push large amounts of investment into the defence industries of southern England, where the case study was conducted. But, though proximity was necessary for the creation of embedded business ties and local growth in these case studies, it was not a sufficient condition. This point was made forcefully by the case study of Asian businesswomen in the UK (*Hardill, Raghuram and Strange*). That study showed strongly that the connections of these businesswomen were with an imaginary and emotional 'home' extending beyond the UK, not a 'home' based on juxtaposition and proximity in a narrow geographical sense.

The bulk of the findings of the case studies in this volume, however, call into question many of the basic tenets of the embedded enterprise model of local growth on the grounds of its inadequate treatment of unequal power relationships, its simplistic treatment of 'institutional thickness' and issues of governance, and its weak conceptualisation of the impact of the role of time and path dependence.

Fundamentally, all the variants of the embeddedness model, and the 'learning regions' model in particular, treat as benign both inter- and intra-firm relationships, and also the relationships between capital and labour. The model paints a picture of growth occurring where collaborative equality serves to mobilise tacit and codified knowledge to generate processes of learning and innovation and, by simple linear extrapolation, growth. What the case studies demonstrate is that in concrete situations, inter- and intra-firm relationships involve, in almost every instance, the exercise of unequal power to exclude, exploit and control network relationships. In developed as much as in developing countries, they identify the essential driver of the capitalist system as the pursuit by the more powerful of profits and the extraction of monopoly rents from those with no capacity or position to fight back. In this respect, they elaborate a point that has previously been made forcefully by Hudson (1999) that the model fundamentally misreads the workings of capitalist economies.

The power to exclude was most clearly evident in the cases of Fiji business (*Taylor*) and Israeli Arab entrepreneurs (*Sofer and Schnell*). The power to exclude was exercised in Fiji by foreign firms and TNCs to avoid risk and to maintain rates of return set outside the country by corporate parents. To all intents and purposes, a variant of the same power was used by large Indo-Fijian businesses to secure their own continuing access to politically controlled commercial opportunities. In Israel, exclusion was a power exercised by Jewish business, leaving Israeli Arab businesses 'over-embedded' in the Arab community and vulnerable to potentially stultifying social pressures within that community. These two case studies also showed clearly that business-to-business structural power can not be divorced from other dimensions of social, cultural and political power. They are all intimately intertwined. Also, the social and economic divisions in the communities in these two case studies meant, in effect, that the conditions that might have created social capital in those places was severely constrained, if not destroyed. 'Exclusion' in those disadvantaged communities brought in its wake embeddedness as a coping mechanism – a way of subsisting in business that was no better than working for wages. Indeed, the 'coping' associated with powerlessness and poverty in developing countries, was also made forcefully by *Mohan*.

The power to exploit was evident in several of the case studies, and most explicitly in *Ekinsmyth's* study of freelance working in magazine publishing in London. In this industry, the appearance of embedded network ties is strong. Friendship, word-of-mouth recommendation and proximity are the underpinnings of a project mode of production that is typical of the burgeoning cultural industries, including advertising and popular music (Grabher, 2001). As in *Hardill, Raghuram and Strange's* analysis of Asian women entrepreneurs, here are strong imagined relations with a community that might almost be called a 'professional home'. But, what has been created is not a benevolent and benign system of collaboration, but a form of self-exploitation that last found its full flowering in the factory-outworker systems of nineteenth century manufacturing, when it was condemned in Parliamentary enquiries into sweated labour for causing the breakdown of family life in places like the West Midlands. The current variant of self-exploitation is just as much a treadmill, and one that forces people to ratchet up their own work rates in a way that Taylorism was never able to do. This same self-exploitation of labour born of enterprise embeddedness finds resonance in *Openshaw and Taylor's* exploration of UK electronics subcontractors, where agency workers and homeworkers are used as mechanisms to bring flexibility to the labour market and to facilitate the avoidance of sunk costs in business.

The power to control business relationships was clearly demonstrated in *Wølneberg's* analysis of the workings of the leather industry supply chain as they impact on tanneries in Argentina. The control exercised by buyers, especially those in Europe and North America, was to the advantage of large-scale enterprises in Argentina, but even these organisations were vulnerable to shifting buyer demands. In addition, national and international regulatory regimes served strongly to reinforce the processes of control and domination that ran through supply chain relationships, further constraining and undermining locally embedded business relationships within Argentina, preventing the creation of social capital. The same processes of control within supply chains was evident in *Hayward, Stringer and Le Heron's* study of agriculture and horticulture in New Zealand, with the implication that local shifts in production are locally indeterminate. *Taylor's* report of experiences in garment production in Fiji reinforces these conclusions but on evidence from a very different buyer-driven commodity chain. It also shows the constraints on 'learning' and the dissemination of knowledge and technology within commodity and supply chain relationships, limiting the creation of locally embedded inter-firm ties and social capital.

An important thread running through the critique of embeddedness developed in the chapters of this volume concerns the institutional structures of regulation and governance that are theorised as the 'supportive tissue' of the collaborative mechanisms of learning, innovation and local growth. In Turkey (*Eraydin*), the institutions of government and civil society have created the appearance of embedded networks among local firms, together with the apparent processes of learning and information exchange that have been said to create self-sustaining local economic growth. But, recession and the almost immediate retreat to cost-cutting strategies, proved these institutions and policies to have created only the 'accidents' of embeddedness – the superficial trappings – and not the 'substance'. The institutions of labour market regulation associated with skills training in London (*Leonard*) demonstrated a similar mismatch between regulation, 'institutional thickness' and local growth. In Argentina (*Wølneberg*), not only were local institutions ineffective in generating self-sustaining local growth, but also the actions of international institutions (including the World Bank and the EU) tended to erode any benefits that might have been created by local institutional support. In combination, the analyses of this volume suggest that 'institutional thickness' is perhaps the weakest stylised fact to have emerged from theoretical institutionalist perspectives on local economies. Indeed, a picture is painted of a frustratingly imprecise concept, shrouded in a fog of contingency that is drained of most of its explanatory power.

Finally, the contributed chapters highlight the inadequate incorporation of time into the processes of the embedded enterprise model. At the heart of the model is what might be called the 'institutional instantaneous' – the atemporal translation of socially networked inter-firm relationships into structures of instant knowledge mobilisation and exchange, learning, innovation and social capital formation. Then, principally by inference, these mechanisms generate immediate local growth. The 'institutional instantaneous' is, in this sense, the equivalent of the assumption of the perfect mobility of capital in economics. Its effect, however, is to all but deny the path dependent, sequential development of socially constructed economies, notwithstanding the significance of those mechanisms as they have been recognised in evolutionary economics.

The case studies present a very different picture of the impact of time on embeddedness. This picture is one of shifting conditions, especially regulatory conditions, changing the foundations on which network relationships are built (*Openshaw and Taylor, Leonard, Eraydin, Wølneberg, Hayward, Stringer and Le Heron*). As a result, observed relationships are frequently interpreted as anachronistic, and the relics of past conditions that are being progressively eroded. These case studies reinforce Staber's (1996) criticism of studies of industrial districts in general and the Third Italy in particular, that they are cases selected on the dependent variable: they are cases chosen for study because of their past 'success', and the processes measured now are reckoned to account for that 'success'.

The simplistic treatment of time in the embeddedness model is both a major theoretical and a major methodological problem. Although the model envisages collaboration and knowledge exchange intensifying in local economies, as tacit knowledge is mobilised, learning deepens and institutional thickness builds, it gives no indication of when lock-in, ossification and institutional overload might arise. These problems are recognised as threats to local growth but they remain largely untheorised (with the possible exceptions of Glasmeier 1991, Grabher 1993). In this volume, *Boschma, Lambooy and Schutjens* begin in a more formal, theoretical way to unpack the issue of time in the context of innovation, but its treatment remains a significant gap in the model. Methodologically, the issue of time is a significant research problem. Qualitative ethnographies of businesses have all the problems of 'survivor surveys' combined with the *ex post* rationalisation of successful survivors. How well the information from these sources can reflect on past and present processes is open to debate – and this is a debate that has yet to begin.

The embedded enterprise model of local growth is an idea that has wide appeal to policy makers and academics alike for very obvious reasons. Its appeal arises because of the response to globalisation it offers to practitioners and analysts; a response which suggests that transnational corporations are not all-powerful and that, whilst the nation state is being hollowed out, it is still possible to act locally while being forced to think globally. Without this kind of perspective we are reduced to a functionalist response to the workings and impacts of global industrial capital. Just as urban analysts sought 'community' in the face of apparent urban 'alienation', so development analysts and practitioners have sought 'local embeddedness' in the face of global corporate authority. But, abstracted from the complexity of reality and the importance of 'context', the concept of embeddedness has been left over-generalised and partial. The principal problem with this embedded enterprise model of local growth, in the various guises in which it appears, is that it is a mechanism that has become a model that has become a mantra. Now, its strength comes from repetition while its weaknesses are ignored. It can be questioned whether it has advanced our understanding of processes of local economic growth and change outside a few cherished localities. It has certainly deflected effort from enquiry into other, less inclusive and more divisive processes shaping capitalist economies.

References

Glasmeier, A. (1991), 'Technological discontinuities and flexible production networks: The case of Switzerland and the world watch industry', *Research Policy*, vol. 20, pp. 469-485.

Grabher, G. (1993), 'Rediscovering the social in the economics of interfirm relations', in G. Grabher (ed), *The Embedded Firm: On the Socioeconomics of Industrial Networks*, Routledge, London and New York, pp. 1-31.

Grabher, G. (2001), 'Ecologies of creativity: the Village, the Group, and the heterarchic organisation of the British advertising industry', *Environment and Planning A*, vol. 33, pp. 351-374.

Hudson, R. (1999), 'The learning economy, the learning firm and the learning region: A sympathetic critique of the limits of learning', *European Urban and Regional Studies*, vol. 6, no. 1, pp. 59-72.

Staber, U. (1996), 'Accounting for differences in the performance of industrial districts', *International Journal of Urban and Regional Research*, vol. 20, no. 2, pp. 299-316.

Index

For Product Safety Concerns and Information please contact our EU representative GPSR@taylorandfrancis.com Taylor & Francis Verlag GmbH, Kaufingerstraße 24, 80331 München, Germany

T - #0093 - 160425 - C0 - 219/152/17 - PB - 9781138734371 - Gloss Lamination